Being, Humanity, and Understanding

# Being, Humanity, and Understanding

Studies in Ancient and
Modern Societies

G. E. R. LLOYD

OXFORD
UNIVERSITY PRESS

# OXFORD
## UNIVERSITY PRESS

Great Clarendon Street, Oxford OX2 6DP,
United Kingdom

Oxford University Press is a department of the University of Oxford.
It furthers the University's objective of excellence in research, scholarship,
and education by publishing worldwide. Oxford is a registered trade mark of
Oxford University press in the UK and in certain other countries

First Edition published in 2012

Impression: 1

British Library Cataloguing in Publication Data

Data available

Library of Congress Cataloging in Publication Data

Data available

ISBN 978-0-19-965472-7

Printed in Great Britain by
MPG Books Group, Bodmin and King's Lynn

# CONTENTS

# Introduction

In *Cognitive Variations* (Lloyd 2007), I combined an examination of some recent work in cognitive science with a discussion of historical data, notably from ancient Greece and China, in an endeavour to clarify some of the deep-seated, but in my view often oversimplified, issues between cross-cultural universalists and cultural relativists on the unity and diversity of human cognition. In this study I use the same tactic, of juxtaposing the investigation of the ancient and the modern world, in an effort to throw light on the extraordinarily rich, but at the same time profoundly puzzling, variety of what we may call cosmologies that are attested whether in our historical sources or in present-day ethnographic reports. In some instances these views are spelt out and provided with a rationale that we can try to come to terms with. In others they are just taken for granted. In both types of case they often defy comprehension, despite the apparent familiarity of some of their assumed components, the concepts of nature and of culture, for instance.

The ideas in question may be about the world, the environment, about ourselves and our relations with other creatures on earth, about spirits and gods, and not just about the animate world, but also about its relations with the inanimate and on whether there is any clear boundary between the two. Being is not a given (however tempting it may be to assume it is) but a problem, and so too is humanity, that is, what counts as being human and on what grounds, and with what implications for how we should behave. The third main topic on my agenda is understanding itself, both *our* understanding of the world, and more particularly our understanding of other people and of how *they* have understood the world. We all use language to talk about the world, but how should we think of the two as connected? That question, among many others, has been much debated, especially though not exclusively in Western philosophy. My aim is to broaden the scope of the inquiry to encompass also what other peoples, at other places and times, have believed or assumed on all these issues.

The evidence available to ethnographers and to historians differs, of course, though the problems both face are sufficiently similar (I would claim) to justify my engaging in an investigation that spans both domains. Ethnographers enter into dialogue with their subjects and check their reactions to the interpretations those ethnographers offer. Ancient historians cannot do that, but on the other hand often have rich sources of information concerning the changes that occurred, over centuries, in what was believed about the problems I have mentioned, where they can try to investigate why such changes took place.

In both history and ethnography we have severe hermeneutic problems in coming to terms with how the world has been understood, to the point where the question has been raised whether indeed it is the same world that is being understood. That is an issue that will recur throughout these investigations. The major methodological difficulty, meanwhile, can be expressed in the form of a dilemma. If we apply our own ideas and concepts to these other systems of beliefs, is that not bound to distort them? But how can we avoid using our concepts, for they are the only ones we have?

I shall have quite a lot to say both about how to go about the study and how not to, but the next major question that has to be confronted is: to what end? What purpose does it serve to wrestle with strange and exotic ideas, even if we *can* begin to understand them? Certainly when the history of science first began, it was studied in order to chart the onward and upward march of human progress. Early ethnography, similarly, was often geared to demonstrating the primitiveness of the societies studied and by contrast the civilized character of the societies to which the ethnographers themselves belonged. Such triumphalist attitudes are no longer acceptable. In science we have more predecessors to draw on, but who can say we are more intelligent than they? With the events of the twentieth century still vivid in our memories, we can certainly not say that we are saner human beings, better able to organize our affairs, than those who used to be looked down on as archaic or primitive. Nor has the twenty-first century begun exactly auspiciously.

But if science can be said to have moved on, why study the endeavours of the past, or those of groups who do not participate in the modern industrialized scientific enterprise, for all that they may have been highly intelligent people who were responsible? There are two parts to the answer. First, where the physical world is concerned, it is far from the case that the whole of ancient, or of modern pre-literate, knowledge can be dismissed as merely archaic, superseded, and errone-ous. Many peoples have achieved deep understanding of their natural environment, and sophisticated mathematics and astronomy are, as we shall see, to be found in ancient civilizations already.

Secondly, and more importantly, every society poses, for us, the problems not just of what its members came up with, by way of understandings of the world around them, but also of how they saw their own place in that world, their humanity in other words, and how they should live their lives. These are difficult questions and we often have to admit that clear answers are not forthcoming. Even more problematic are issues to do with the extent to which beliefs are subject to change and modification, at least in those societies where there is diachronic evidence we can use to address the question. How far were traditional ideas and practices, and the authorities who perpetuated them, open to challenge and in what circumstances did that happen? We have ample testimony, for ancient societies, concerning the impact of political, religious, ideological, and institu-tional factors on the investigations that were pursued, but similar influences are at work on our own scientific and other inquiries today, different though they are in many respects from those of our predecessors. To cite just the most banal example, do not modern authorities insist on results that they assess often in the most simplis-tic materialist terms?

I have spoken, by implication, of the history of ancient 'science', but a further important question that will concern me is what we mean by that term. I have argued, on other occasions (Lloyd 2009), against the narrow, admittedly conventional, view that has it that we cannot properly talk of science before the seventeenth century when it was a development that originated in, and was for long confined to, Europe. Throughout these studies I advocate and adopt a far broader conception, closer to the French 'science' or German 'Wissenschaft' than to our own English usage when we have the modern natural sciences in mind as the paradigmatic, if not indeed the sole, claimants to title. Science, here, I take to cover every attempt systematically to investigate the phenomena, to observe, classify, predict, and explain them, in short to increase understanding of them. Such ambitions can be found in plenty of societies that do not rate a mention in standard histories of world science.

We must allow, as I said, that ideas about what is to count as 'understanding' have varied, and we must be careful not to prejudge how these came to be formulated and justified. The aims of science can be characterized in very general terms, but how the ambition to understand has been embodied and how ideas on that subject have changed pose challenging problems of interpretation that require scrupulously even-handed analysis. This is not to adopt a simple-minded relativism, but it is to allow pluralism on the question of how people at different times and places have gone about the task of understanding and of communicating what it is claimed is understood. I define my remit to include not just the explicit inquiries of ancient civilizations such as Greece and China, but also the often implicit cosmologies recorded in the ethnographic literature. As already noted, the nature of the evidence available differs, and being no anthropologist myself, I limit myself there to some philosophical reflections on what is reported in that literature. But the problems I have identified recur in both ancient and modern societies. They are worldwide and not only can, but should, be addressed ecumenically.

I shall structure my discussion in five parts. The first uses both ethnography and ancient history to tackle the central problems of the understanding of what it is to be human and the ontological assumptions that answers to that question may presuppose. How can we evaluate strikingly divergent views of the relations between humans and the world, comprising animals, nature in general, and even the gods—how can we even begin to understand them? The methodological issues, as noted, are severe and I underline the need for caution with regard to any claims to have fully comprehended others' systems of beliefs, their values and notions of agency, and ways of interacting with their environment. However, that should not prevent us from using the evidence available to us precisely to extend our knowledge about what humans have believed. We may not be able to understand jaguars as others understand them, but we can begin to understand the significance of what people have believed about jaguars for their ideas both about the world and about how to live.

The second chapter takes up a problem broached already in the first, namely how we can account for error and indeed diagnose it without prejudice. We need, to be sure, to engage sympathetically with others' views, but that does not mean that they were or are infallible any more than we are. We are not the only society where doubts and disbeliefs are expressed about the particular and the general views of individuals and of groups. Evidently we need something of a taxonomy of possible modalities of error and some ground rules for their diagnosis.

The third chapter focuses on being in general. It begins by charting the radical changes that the evaluation of ancient science has undergone in recent years, both in the assessment of its strengths and weaknesses and about why it is worth studying. Antiquity provides a precious opportunity to study how alternative cosmologies came to be proposed within a single society, the circumstances in which that can happen, and the extent of the penetration of such new ideas within the society in question. My aim here is to use both ancient and modern materials to throw light on my strategic question of the comparative analysis of ontologies.

Chapter 4 confronts a series of issues to do with the language in which beliefs have been expressed, and the audiences their proposers were endeavouring to persuade. How far and in what ways did the communication situation of those exchanges influence the content of what was proposed and its acceptability? I explore the problems both in relation to ancient Greece and China and with regard to the category of 'myth' around which much anthropological theorizing has revolved. I end with a tentative but controversial proposal to get round the customary Western dichotomies of the literal and the metaphorical, and the mythical and the rational. This is by way of the notion of semantic stretch, which provides an alternative basis for the analysis of language use and so opens up the possibility of a more complex account to do justice to the open-endedness of the discourses in which divergent ontologies are expressed.

The fifth and final chapter seeks to pinpoint the ways the preceding inquiries can be brought to bear on current philosophical problems. How can the combined study of ethnography and ancient science contribute to our understanding of such issues as the debate between realism and relativism, the problems of truth and objectivity, and of the supposed incommensurabilities between conceptual systems and the worlds they describe, where in each case clarifications are badly needed? Thus neither realism nor relativism is a single well-defined position, and the idea that we have to choose between them should be resisted. Again the postulate of incommensurability comes in very different strengths, and it is only at the extreme that it poses the threat of total mutual unintelligibility, cutting human beings off from one another. What can the study of diverging ontologies tell us about human understanding, about the hazards it faces and the opportunities it presents? What can we learn from the encounter that is relevant to our own self-understanding and indeed to the values by which we should live? To understand ourselves we need to understand one another. It is my aim to use the latter to throw some light on the former.

A brief epilogue concludes with a review of what we may be said to have learnt concerning the investigation of the triad of concepts that gives this book its title, namely Being, Humanity, and Understanding. I am well aware of the enormous literature on aspects of the problems that has been produced by philosophers, anthropologists, historians, linguists, cognitive scientists, developmental psychologists, and others, and I shall duly draw, admittedly highly selectively, on this work. Yet I believe that the issues have often been dealt with too much from within the traditional problematic of one or other conventionally defined academic discipline. I recognize how difficult and how risky it is to attempt to combine insights from different approaches dealing with disparate data, adopting different criteria for success and practising different methodologies. But that is the programme that I have ambitiously set myself in this study.

I have had the great good fortune to be able to consult many friends and colleagues from a wide variety of disciplines. My earlier study, Lloyd 2007, together with an article based on it, Lloyd 2010b, was the subject of a series of critical contributions in the *Interdisciplinary Science Reviews* in 2010 and I wish to thank the editors, Professors Willard McCarty and Brad Inwood, for organizing this, and the contributors for their perceptive comments. In addition, I have had sustained conversations and correspondence with several scholars who have generously allowed me to try out my ideas, among whom I may mention Anne Cheng, Lorraine Daston, Philippe Descola, Rob Foley, Martin Holbraad, Stephen Hugh-Jones, Nicholas Humphrey, Bruno Latour, James Leach, Amiria Salmond, Simon Schaffer, Roger Smith, Marilyn Strathern, Aparecida Vilaça, and Eduardo Viveiros de Castro especially, not that any of them should, of course, be held responsible for how I have used their advice. I have also learnt much from the constructive comments made by two anonymous readers for Oxford University Press. Chapter 1 of this book is based on ideas that I presented in my 2010 Henry Myers lecture, and chapters 3 and 5 similarly develop arguments from lectures that I delivered at the Collège de France and at the University of Zaragoza during that year. Sections of my discussion were also later presented at the National University of Singapore and at City University, Hong Kong. I should like to give my warmest thanks to Hilary Callan, Roy Ellen, Anne Cheng, Juan Vicente Mayoral de Lucas, Gregory Clancey, and Zhang Longxi for their invitations and to my audiences for their comments. Finally, it is a pleasure once again to thank Peter Momtchiloff and his colleagues at Oxford University Press for their support, encouragement, and professionalism.

*GERL*
*July 2011*

# 1

# Humanity between gods and beasts?

ONE of the greatest modern interpreters of ancient Greek thought, who was one of my mentors as well as a close friend, Jean-Pierre Vernant, together with his colleagues Marcel Detienne and Pierre Vidal-Naquet, developed an influential argument according to which for the ancient Greeks, what it is to be human was defined by a triadic relationship, in which humans are sandwiched between gods on the one hand, and the other animals on the other.[1] That might seem just an idiosyncratic fantasy of some ancient Greeks who are notorious for their vivid anthropomorphic imaginations. But in this chapter I set out to explore the general issue, namely how did, how do, other societies get on with the business of defining what it is to be human or what assumptions do they make on that score? Is the question of what is distinctive about humans perceived as a problem, or is the concern not with other animals versus us, but with the differences between different human groups, between us and others, outsiders, strangers, enemies, who may or may not be humans exactly like us?

The question of whether there are robust cross-cultural universals in this area, that is, in the understanding of what it is to be human, has recently received a lot of attention in anthropological debate, stimulated by the work of Viveiros de Castro and Descola especially. My

---

[1] This notion of the triadic relationship was very much the product of the work of Vernant and his colleagues in the 1970s. In 1972 Vernant wrote the introduction to Detienne's *Les Jardins d'Adonis*, which was given the title 'Between the beasts and the gods' when the translation was reprinted as ch. 7 in Vernant 1980. In the same year in which Detienne published that book he wrote an article entitled 'Entre bêtes et dieux' in *Nouvelle revue de psychanalyse* 6 (1972) 231–46, later reprinted in a revised and retitled version in Detienne 1977 and translated into English in 1979. Vernant's other colleague, Vidal-Naquet, wrote an article entitled 'Bêtes, hommes et dieux chez les Grecs' for a colloquium in 1973, later published in 1975.

own studies of ancient societies, China as well as Greece, allow me to bring to bear some non-Western as well as pre-modern materials on such subjects. And since the question of alternative ontologies, alternative worlds, is in part a philosophical problem, I shall venture some comments on that too from my position on the sidelines.

In fact, since the nineteenth century, attempts to settle on the defining characteristics of what it is to be a human have succeeded one another in Western thought with unnerving rapidity.[2] No sooner had one proposal been tried and found wanting than another would be put forward, for it, in turn, more often than not, to founder. Is language an adequate criterion? That used to be one of the favourites. We find Aristotle already distinguishing voice (that he recognizes in birds, for instance) from speech (*History of Animals* 536b1). But animals do not just communicate general messages: evidently bird song is used to claim a territory and find a mate, but there is much that we do not understand about the *virtuosity* of some bird performances: what does that add? There are still many questions we do not yet know how to answer, as such authorities as Catchpole and Slater (2008) recognize. But animals also convey *specific* information. Vervets, for example, have one cry when the danger is a snake, another when it is a panther, a third when it is an eagle.[3] And other creatures in their habitat can interpret those cries correctly: they understand the vervets' language, we could say, as is also the case when the vervets interpret the alarm calls of other species, such as starlings, correctly. That is behaviour acquired in the wild, and in that respect is more impressive than the well-known abilities of apes, such as the famous Washoe and Sarah, to recognize and manipulate signs, to solve puzzles, to express wants and feelings, and to attribute intentionality to others, for in those cases they were taught, of course, by humans. They have sometimes been cited, indeed, as evidence for

---

[2] Smith 2007 outlines the trajectory of European thought on the subject of human nature. Leach 1972 (cf. 1982: ch. 3) contributed a characteristically provocative study of humanity and animality, as did Ingold 2002, and Agamben 2004 is another subtle writer who has persisted in the quest for an answer to their difference. Most recently, the summer 2009 issue of *Daedalus* was devoted to the question of what it is to be human.

[3] See Seyfarth and Cheney 1982, Cheney and Seyfarth 1990. Understanding the alarm calls of other species is not just a gift of vervets, of course. Avital and Jablonka 2000: 246, for instance, discuss how meerkats respond to the signals of ground squirrels.

the limitations of animal language acquisition, rather than for its potentialities.[4]

Another favourite was tool-making: homo is homo faber. But by now the evidence for some animals' abilities in that regard is overwhelming. Where there used to be just a few standard examples of animals using stones as hammers or anvils, we now have detailed documentation of, for instance, the techniques that chimpanzees have devised for ant or termite fishing.[5] Then there are also studies that show that some individual members of certain species are quick to learn from their conspecifics while others are very slow to do so. But if not all species members show the same skills, that suggests a prima facie contrast between behaviour that is acquired, and what is innate, instinctive, or as we now say hard-wired, even though, as Bateson and Mameli (2007) and Nettle (2009) have recently insisted, that dichotomy, between the innate and the acquired, is less clear and less clear-cut than is often assumed.

If language and technology turn out not to be the decisive discriminators they were once thought to be, what about symbolic representations, the ability to picture others' thoughts, to have a model of the mind, as the philosophers put it, the key, surely, to the human development of culture?[6] Once again, what seemed firm boundaries to earlier generations of researchers have been eroded. Once again, the primatologists have led the way. In groups whose survival depends on close cooperation between the whole community, cheater detection mechanisms are needed to control or at least restrict free-riders, and these have been studied in a variety of species.[7] Nor is it just primates that have complex social organizations. Where bees used to be the favourite example of this, with their social differentiations and communication abilities (the famous 'dance'), ants are now just as often cited, notably for their ability to 'farm' aphids as a source of food, and

---

[4] Gardner, Gardner and van Cantfort, 1989, recorded Washoe's feats, and Premack 1976 and Premack and Premack 1983 those of Sarah. For a sceptical view about these as examples of language, see Pinker 1994.

[5] See Boesch and Boesch 1984 (cf. Kummer 1995) on the use of stone anvils and hammers, and Boesch 1996 and Humle and Matsuzawa 2002 on ant-fishing. McGrew 1992 gives an overview of the literature to that date. For non-human animals in general, see Griffin 1984, 1992.

[6] See Byrne and Whiten 1988, Boyd and Richerson 2005, Richerson and Boyd 2005, Dunbar 2008.

[7] See especially Cosmides and Tooby 1992, cf. Dunbar 1999.

the habit, well known already to Darwin, of some species of ants to enslave other species.[8]

It was indeed Darwin's work especially that triggered the heightened interest in the question of the defining characteristics of humans in the mid-nineteenth century. It seemed to many more and more urgent to be able to draw the line between us and the other animals. Some were out to fill in the details of evolutionary theory, while others, to be sure, sought to gather evidence to refute it. But the question itself had, of course, a massively long history centuries before Darwin, in which Christianity looms large. *Genesis* had already taught that God created humans in his own image, giving them dominion over all other creatures. When they die, they just die: but we, as the *New Testament* and the subsequent teaching of the Church emphasized, are destined for resurrection, when eternal bliss in heaven, or damnation in hell, awaits us.

Yet the Christian belief in the afterlife certainly did not and does not resolve all the questions, not even for Christians. One that could and did repeatedly arise was whether creatures that certainly looked like humans were indeed endowed with immortal souls. When the conquistadors encountered native Americans in the Antilles, they wondered whether they were truly humans with souls that were there to be saved. The native Americans, for their part, conversely, worried whether the Spaniards had the same *bodies* as they themselves did. They drowned some of their Spanish prisoners to see if their bodies *rotted*. Lévi-Strauss, when citing this incident, pointed to the asymmetry: the Spaniards were tackling the problem like social scientists, while the Americans adopted what could be called a biological, scientific approach.[9] Viveiros de Castro has also cited that encounter in connection with his ideas on perspectivism, which I shall be considering shortly, but I use it now just to show that the Christian dogma still left open the question of *which* creatures were truly humans, endowed with those immortal souls.

But many other answers besides Christian ones were given in antiquity to the question of what differentiates humans from other animals. The Christian legacy had, to be sure, particular impact on

---

[8] Darwin 1859 ch. 7.
[9] Lévi-Strauss 1955/1973: 76, 1973/1976: ch. 18. Cf. Viveiros de Castro 1998: 475, 478–9 and 2009: 14f., Descola 2005: 386, Latour 2009.

later European beliefs, but the issue was not just a concern for Christians, nor (as we shall see) just for Western thought.[10] The Greeks, with whom I started, did not all agree on the subject—they hardly ever did. For the majority who practised the city-state religion, the distinctions that Vernant drew attention to were the ones that counted. The gods ate nectar and ambrosia and did not die. Beasts did not eat bread or cooked food and some of them were appropriate offerings for humans to make to gods. Humans in between ate bread, died, and entered into communication with the gods by means of sacrifices, of animals, or of incense, or cakes. We have here then a classic ancient statement of the view that you are what you eat.

Such beliefs about the gulf between humans and other animals were lived out and confirmed in state and domestic rituals throughout the year and were the subject of countless myths. But the philosophers came up with other ideas (as they often tend to do). Let me mention just three, those of Empedocles, Plato, and Aristotle. Before Empedocles, Pythagoras had already taught the doctrine of metempsychosis, the transmigration of souls, which had the important moral consequence that how you were reborn—into what kind of animal or even plant—depended on how you lived your current life. But while Pythagoras himself is a rather shadowy figure, we know from the extant remains of the poetry of Empedocles how he developed the notion. Transmigration already implied that all living things are kin—thereby eroding the contrasts between them. Empedocles spelt out the implication, that it was wrong to take life, wrong, therefore, to sacrifice animals on the altars of the gods.[11] That overturned one of the cardinal institutions of the city-state religion, and remarkably, unlike some other religious innovators, he got away with it.

Plato in turn flirted with the idea of transmigration in the *Republic* and *Timaeus* especially: it suited his moral teaching nicely that there were rewards and punishments in the afterlife. That underlined Socrates' message that the most important goal in life was to care for your immortal soul. But Plato's most influential doctrine related rather to

[10] Sahlins, who did a brilliant job of exposing the perversity of Western ideas on human nature (Sahlins 2008), might have spread his net wider to encompass analogous preconceptions in, for instance, Chinese thought.

[11] Empedocles Frr. 136, 137.

his notion of a tripartite soul. The appetitive and the spirited parts, which humans share with other animals, should be under the control of the rational faculty, the *logistikon*, which in the *Timaeus* (though not, it is true, so clearly in other dialogues) alone is immortal.[12] This is very much a human faculty, and in the fanciful account of the generation of other creatures which we are given at the end of that dialogue, they arise when they lose their reason, *nous* (*Timaeus* 92bc).

Aristotle knew a lot about animals, more than any other ancient Greek, but he too drove a wedge between humans and them by way of the reasoning faculty, though there are considerable complications. First, he recognized plenty of cognitive faculties in animals, not just perception but also imagination, intelligence (*phronēsis*) including technical skills and the ability to learn.[13] When he set out to consider the differences between all animals, humans included, he drew distinctions between their manners of life (*bioi*), activities (*praxeis*) and characters (*ēthē*) as well as in their physical features, their parts (*moria*) (*History of Animals* 487a11ff.). Some animals know what herbs will help to heal their wounds (612a1ff.), one species of spider has a sense of the geometrical centre of their webs (623a7ff.), the construction techniques of swallows when they build their nests resemble those of human builders (612b18ff.), while the complex social organization of bees directed at the common good of the hive excites his admiration (623b17ff., 625b17ff., 627a20ff.).

None of that stopped him from insisting, in general terms, that it is reason that marks out humans. But the second complication is that when reason, *nous*, is said to be immortal, this is not what he calls the passive reason, which ceases when we die, but the so-called active reason, where it is not at all clear whether he is talking about something human or something exclusively divine.[14] God, for Aristotle, is

---

[12] The key passage is Plato, *Timaeus* 69c–72d. Whether the Divine Craftsman or Demiurge who is responsible for creating the cosmos is to be identified with *nous* (reason) is a famous crux. But his achievements are certainly called the works of reason, *Timaeus* 47e.

[13] Lloyd (forthcoming) sets out and comments on Aristotle's views on the cognitive capacities of animals other than humans. While his official theory is that other animals do not share in moral or intellectual reasoning, in practice he allows them many other activities that involve calculation and learning from experience. Moreover, like most ancient Greeks, for whom the lion is the paradigm of courage, the deer of cowardice and so on, Aristotle saw many analogies between human characters and those exemplified in other animals.

[14] The only text in which Aristotle draws a distinction between two types of reason is *On the Soul* III ch. 5, 430a10ff., 24f., the subject of extensive commentaries, and unresolved debate, in antiquity as well as in modern times.

identified as an Unmoved Mover who is responsible for the movements of the heavenly bodies not by pushing them, but by being the
object of their love (so he is a final cause, then, the good they aim at,
not an ordinary efficient cause). That suggests the remoteness of the
divine, but if we ask what activity God engages in (for he can hardly be
idle or asleep), the answer is that he uses his *nous*, not in practical
reasoning nor in deliberation, but in intellectual activity (*theōrein*,
*theōrētikē*) which is also within our human reach. However, since the
object of God's thought must be the best thing of all, it turns out to be
himself, and so his thinking is, in the striking phrase of *Metaphysics*
1074b34f. (cf. 1072b18ff.), a thinking of thinking, *noēsis noēseōs*.
While God's thinking is continuous and eternal, we are only able to
engage in intellectual activity—the study of intelligible forms—from
time to time (*Metaphysics* 1072b24ff., 1075a7ff.). Yet such a study,
in Aristotle's view, is the highest faculty of which we humans are
capable.[15]

So once again, we have a sandwich. God is the highest form of
reasoning: he does not need to perceive nor imagine nor move nor
even deliberate, let alone reproduce. We share intellectual activity
with God, but our reasoning is also practical. Moreover, we share the
other vital faculties, perception and the rest, with the other animals,
though they lack our intellectual and deliberative capacities (*Nicomachean Ethics* 1178b27f., *History of Animals* 488b24ff.). But this is not
just psychology, but morality, for we are enjoined to imitate God's
activities as far as possible, to immortalize ourselves as far as possible,
*eph' hoson endechetai athanatizein*—thus contradicting the usual
Greek advice, that as mortals we had better have merely mortal
thoughts or aspirations, *thnēta phronein*.[16]

Strange as the notion of an Unmoved Mover is, Aristotle's holding
to immortality and his privileging reasoning as distinctively human
seem familiar enough. It is salutary, therefore, to contrast a very
different ancient culture—namely China—with its distinctive ideas
about the gods, religious practices, and philosophical speculations.
Among the latter we find a notion of something like a Scala Naturae
in the third century BCE writer Xunzi. Water and fire, he says, have *qi*
(that spans breath and energy) alone; grasses and trees also have life

---

[15] Aristotle, *Nicomachean Ethics* 1178b7–28.
[16] Aristotle, *Nicomachean Ethics* 1177b31–4.

(*sheng*) to which birds and beasts add knowledge (*zhi*). But humans have *qi*, and life, and knowledge but also an extra faculty, which makes us the noblest of them all. This turns out to be not the reasoning faculty as in Aristotle, but righteousness, *yi*, that is, a *moral* capacity.[17]

There are two main points about this view of Xunzi's. First, he does not extend the Scala upwards to gods. Elsewhere he is critical of many of the rites regularly directed to spirits and demons, which he thinks a waste of time and resources.[18] But no more does he, in his Scala, invoke *tian* (heaven), for instance, conceived by many Chinese as an impersonal deity, though it is certainly no transcendent creator-craftsman. So gods, let alone the ancestors to whom most Chinese directed much of their worship, do not come into it.

Secondly, the problem about using morality to differentiate humans is that, as the Chinese like the Greeks were very well aware, human views on right and wrong differ widely. That made it difficult to maintain that all humans share the same sense of what righteousness is, or even the same sense of its importance. The moral sense has often been used to suggest not what all humans share, but what many humans lack. In many societies the boundaries of the human stop at those of the society itself, though that is not just an idea about morality. Many have other names for other groups, but refer to themselves just as 'the people'. There is no need to travel to Amazonia or to Africa to illustrate the point. My own forebears, the Cymru, will do very nicely. That is a comparatively mild form of exclusiveness. But far stronger versions are widely attested. Many collectivities use their enemies the better to define themselves: they need their enemies in that sense. But that leaves open the questions of what kinds of beings those enemies are, where they came from, and whether they are indeed permanent enemies (Viveiros de Castro 1992, Vilaça 2010).

Meanwhile, the converse of the idea that certain humans are not like us—and so not human—is the equally common notion that many non-human animals are persons just as much as we are. That idea may be extended to what we consider inanimate objects, such as winds or

[17] *Xunzi* 9, 16a: Knoblock (1988–94) ii 103–4. The question of whether or in what respects a moral sense can be attributed to certain animals continues, of course, to be hotly disputed. See de Waal 1991 and 1996 on reciprocal altruism in chimpanzees, cf. Seyfarth and Cheney 1984 on vervet monkeys.

[18] *Xunzi* 21, 8: Knoblock (1988–94) iii 108–10.

stone, and even to artefacts, such as kettles. Among many commentators who have explored these ideas, Hallowell (1960) on the Ojibwa and Howell (1984) on the Chewong are exemplary (and cf. the overview in Viveiros de Castro, e.g. 2009: 15f.). The problem of the concept of the person itself was opened up, in modern times, especially by Mauss (1938) (see Carrithers, Collins and Lukes 1985) and more recently has become, if anything, even more controversial. Many societies in Melanesia and Amazonia especially are reported as seeing what we might think of as bounded, stable, individuals as essentially defined by networks of relations. Persons are, on this view, then, not individuals, but rather, in the term made famous by Marilyn Strathern, 'dividuals', divisible into multiple components formed from relations with a plurality of other persons and subject to recombinations in constant disequilibrium (Strathern 1988, 1999, 2005, cf. Wagner 1991, Mosko 2010, Vilaça 2011). Equally, notions of the constitution of groups vary, not just who belongs to a particular actual group, but the distinguishing marks of groups themselves, particularly when these too are relational and the relations are indeed reversible (as recently discussed by Vilaça 2010). I shall have more to say on these anthropological issues in due course.

To be sure, in Western thought since at least the mid-eighteenth century there have been repeated axiomatic assertions of the equality of all human beings. Rousseau famously put it in the *Contrat Social* (1762) that 'man is born free and everywhere is in chains'—a dictum that puts all humans on an equal footing, joined by their common predicament. The American Declaration of Independence (1776) stated: 'we hold these truths to be self-evident, that all men are created equal', though it went on with an invocation of anything but a self-evident notion of a Creator ('that they are endowed by their Creator with certain inalienable rights'). The French Declaration of the Rights of Man (1789) followed suit and the motto *Liberté, Egalité, Fraternité* became the best-known slogan of the French Revolution. By 1948 the United Nations Declaration of Human Rights had as its first article 'all human beings are born free and equal in dignity and rights'. But these are all expressions of an ideal that have to be set against the background of the realities of enormous, exponentially increasing inequalities, in material resources and so also in opportunities for self-fulfilment.

In antiquity, the Chinese mostly despised the alien peoples by whom they were surrounded, just as the Greeks did their barbarians who could not speak Greek and just 'ba-baed' away. The Chinese names for many non-Han tribes incorporate the radicals for animals, especially the dog. But against that, the Chinese term *ren*, like the Greek *anthrōpos*, refers to humans in general, non-Han as well as Han, women as much as men. Xunzi's *yi*, righteousness, is an idealization: but then so was Aristotle's *logistikon*, which he had no compunction in saying was possessed in only a very limited way by women and by slaves.

But Xunzi was not the only Chinese theorist of what makes humans humans. In another third century BCE text, the *Lüshi chunqiu*, we find that while what separates different human groups are such items as languages, customs, dress, taste, what unites them all is feelings, especially the need to satisfy certain basic desires.[19] Speaking of two tribes called the Man and the Yi, the text says that

> despite their backward tongues, their different customs and odd practices, despite their clothes, caps and belts, houses and encampments, boats, carts, vessels and tools, and despite their preferences of sound, sight and flavour all being different from ours, [they] are one with us and the same as us in satisfying their desires.

It is precisely because we all have the same basic desires, physical ones, no doubt, especially, that we are all humans, non-Han and Han, sage kings and tyrants.

So quite different views were expressed within ancient Greece and again within ancient China on the question of what is distinctive about humans. There was no consensus, let alone an orthodoxy, on the subject in either society. But even more radical divergences are reported in the modern ethnographic literature. Viveiros de Castro's perspectivism and Descola's ontologies are two of the most outstanding contributions to the issues and, as I intimated before, I shall concentrate on them.

Viveiros de Castro introduced perspectivism in pioneering papers in 1996 and 1998, a set of reflections, so he tells us, on an indigenous theory extensively reported, but at that stage little commented on, in

---

[19] *Lüshi chunqiu* 19.6.2. Knoblock and Riegel 2000: 497–8.

particular from Amazonia but far from limited to that area.[20] According to this, 'the way humans perceive animals and other subjectivities that inhabit the world—gods, spirits, the dead, inhabitants of other cosmic levels, meteorological phenomena, plants, occasionally even objects and artefacts—differs profoundly from the way in which these beings see humans and see themselves' (1998: 470). He then explained in a famous passage:

> Typically, in normal conditions, humans see humans as humans, animals as animals and spirits (if they see them) as spirits; however animals (predators) and spirits see humans as animals (as prey) to the same extent that animals (as prey) see humans as spirits or as animals (predators). By the same token, animals and spirits see themselves as humans: they perceive themselves as (or become) anthropomorphic beings when they are in their own houses or villages and they experience their own bodies and characteristics in the form of culture—they see their food as human food (jaguars see blood as manioc beer, vultures see the maggots in rotting meat as grilled fish, etc), they see their bodily attributes (fur, feathers, claws, beaks, etc) as body decorations or cultural instruments, they see their social system as organized in the same way as human institutions are (with chiefs, shamans, ceremonies, exogamous moieties, etc). This 'to see as' refers literally to percepts and not analogically to concepts, although in some cases the emphasis is placed more on the categorical rather than on the sensory aspect of the phenomenon. (1998: 470)

Descola's ambitious fourfold schema presents a potentially exhaustive classification of ontologies, differentiated in terms of the notions of

---

[20] Viveiros de Castro 1998: 484 notes 2 and 3 itemizes the large number of groups from Amazonia and elsewhere across the world in which perspectivism is present. There are two possible misunderstandings that centre on Viveiros de Castro's use of the term 'indigenous theory' in relation to perspectivism. Some such as Willerslev, who offers a highly deflationary account of the perspectival phenomena, appear to have taken 'theory' as implying abstract speculation (opposing that to meanings that emerge from 'concrete contexts of practical engagement', Willerslev 2007: 94), but that is no part of Viveiros de Castro's interpretation. Nor is there, in that interpretation, any suggestion that there is uniformity in the lived experiences of the peoples in whom perspectivism is found. On the contrary, he acknowledges that 'people shift ontological gears every minute', though that leaves open the question of which of the presupposed, implicit, everyday context-bound ontologies is 'collectively foregrounded as conceptually worthy of reflection' (pers. comm.).

physicality and interiority they assume (Descola 2005).[21] Physicality relates to what is believed about bodies and physical objects generally: interiority concerns how one feels inside oneself, and beliefs about other persons or other intentional beings. The taxonomy proceeds by way of a contrast, in each case, between a view that stresses continuity, and one that emphasizes discontinuities, between humans and non-humans. Thus on the physicality axis, the view that stresses continuity maintains that all physical bodies are essentially the same: here the opposing one, which emphasizes discontinuities, stresses differences, and postulates that what marks out different kinds of entities is, precisely, the bodies they inhabit. Similarly, on the interiority axis, the emphasis may be on continuity (the souls of all beings are the same) or on the discontinuities (humans are a race apart).

This double differentiation yields a fourfold schema. 'Animism' (not to be confused with Tylor's notion) is now redefined as an ontology in which other beings besides humans have spirits, but what differentiates them is their bodies. Interiority, selfhood, is common: physicality is what differentiates things. It is because the jaguar has the body that it has that it sees things in the way it does. 'Totemism', secondly, again redefined, is now applied specifically to an ontology where particular species of animals, plants or other objects share with particular groups of human beings some complex of essential qualities that differentiates them from all other similar groupings. The unity or continuity between humans and non-humans is assumed both on the physicality axis and on the interiority one. Intentional beings (human or non-human) and physical bodies (again human and non-human) form a continuous, seamless, whole. Thirdly, 'analogism' is the reverse of totemism, since it assumes discontinuities on both axes, postulating differences between humans and non-humans in both their physicality and their interiority, but finding analogies and correspondences across the domains thus differentiated. Descola sees this as the dominant ontology in both ancient Greece and China, for instance,[22] and claims it was more or less universal in

---

[21] Descola 1992 and 1996 had earlier offered a simpler triadic analysis of animism, totemism, and naturalism, before proposing the fourfold schema of Descola 2005 (cf. also Descola 2009).

[22] However, Descola 2005: 287–8 limits himself to only brief remarks on Chinese analogism. I shall be returning to both Greek and Chinese ontological pluralism in chapter 3.

Europe throughout the Middle Ages and the Renaissance. Finally, there is what he calls 'naturalism', the default ontology of modernity, where physicality is unified, but interiority divided and discontinuous. This view insists on differences between humans and non-humans on the interiority axis with respect to minds and cognitive faculties generally (humans alone have culture in the strict sense), but sees humans and non-humans as linked by their shared physicality. We are all made of the same stuff.

There are many aspects of these ideas that I cannot go into here. I have elsewhere questioned how far 'analogism' will do as an over-arching rubric to capture the diversity of views on offer from both Greek and Chinese antiquity,[23] and one major problem for Descola certainly relates to how what he calls 'naturalism' arose from whatever it is imagined it developed from (analogism, principally), for on his account it was nowhere to be found before it came to dominate thinking in seventeenth-century Europe. A further complication, of which Descola is well aware, 2005: ch. 3, might be that one version of naturalism, which maintains providential teleology, appears in that respect to share a notion of divine intentionalities at work in the world that is found in some animist beliefs. In his new book (2009: 48ff.) Viveiros de Castro identifies the fundamental differences between his perspectivism and Descola's animism, over and above the point that it did not form part of a comprehensive taxonomy of ontologies.[24] He insists that what perspectivism offers is not an elaboration or a type or subspecies of animism, but rather a concept, a theory, the basis of an indigenous metaphysics. It is as much an anthropology as the conceptual apparatus of the anthropologists who study it, a counter-anthropology, in other words, a metaphysics of cannibalism

---

[23] Lloyd 2007: 144. In practice, Descola 2005 adds many qualifications to the fourfold schema when he engages in a detailed examination of their historical manifestations, in particular with regard to the different relational modes (predation, exchange, gift, transmission, and so on) with which they are associated.

[24] Some of the same Amazonian data receive very different analyses in the two interpretations. Descola's animist reading brings to bear his usual contrast between physicality and interiority—the body is a set of physical dispositions that predispose an animal to certain kinds of behaviour, while the common interiority shared by different creatures allows communication between them. But Viveiros de Castro takes the contrast between body and soul to be one rather between figure and ground. The notion of the body, for many Amazonian peoples, he says (Viveiros de Castro 1998: 478 and 481) already incorporates affects and memories (cf. Vilaça 2010 on the Wari' concept of *kwere-*).

indeed. But setting aside the very considerable subtleties and complexities, I shall focus here on my question of ideas on the relations between humans and other beings.

On that question, the implications of both these—very different—contributions are clearly momentous. On Viveiros de Castro's account, humans turn out to be quite different when seen by spirits or by animals (and that will differ again depending on whether those animals are human predators or human prey: and they may be both. The contexts change and they may be all-important). But animals and spirits see themselves as humans: so what humans and animals have in common is not animality (as in Aristotle and Xunzi) but humanity. In Descola's schema we have four distinct answers corresponding to the ontology in play, where humans are distinguished from other creatures either by interiority alone, or by physicality, or by both, or by neither. Whether we are or are not the same kinds of beings as other animals, and if so, why, and if not, why not, are questions that receive radically different answers depending on the ontology adopted.[25]

A first conventional reaction to this might be flatly to deny the possibility of these alternative worlds. On one view of what we mean by 'world' there can only be one, the universe as the totality of what exists, and we know how to investigate that, namely, with what we call science. But that would be to miss the point. What is suggested here is that within this—single—universe, different beings, different animals, and also different members of the human race have such different experiences, perceptions, and ways of interacting with their environment that we should think of them as living in different worlds.[26] The thought would then be that conscious beings construct their world as they interact with it, although that is not to say that any such construction is purely arbitrary.

Something of a digression is needed to attempt a first clarification of the philosophical question of the possibility of alternative ontologies

---

[25] Viveiros de Castro 2009: ch. 3 elaborates the concept of 'multinaturalism' in the context of a contrast between it (and the perspectivism it implies) and the mononaturalism, but multiculturalism, of the dominant contemporary Western view.

[26] Cf. Sperber 1985: ch. 2, who explores the possibilities in terms of what he calls 'cognizable worlds'. Cf. also Goodman 1978, Ingold 2000, Henare, Holbraad and Wastel 2007. Viveiros de Castro 1998: 477f. put it that all living beings see the world 'in the same way': what changes is the world that they see. The things that they see are different: what to us is blood is maize beer to the jaguar.

and for that I shall use an excursus on Greek and Chinese thought. At first sight the usual Greek view about the constitution of physical objects seems familiar enough. The furniture of the world consists principally of solid, discrete, impenetrable objects that fall if they are not supported, that move in response to collision with other such objects, and so on. Some cognitive scientists indeed would claim that all humans have a domain-specific cognitive module that yields such an intuitive ontology, though everyone has to concede that modern chemical understanding complicates the picture and that is before we come to the paradoxes of fundamental particle physics. Yet to that last point the cognitive universalists would counter that those are just theories and do not correspond to what experiments can reveal to be universal constraints on human cognition. But even if it were shown that children everywhere assume the same naïve physics, they—we— have to grow out of that, just as we have to grow out of naïve psychology (see, for example, Gelman and Spelke 1981), and indeed we do so in different ways, with differing views as adults, as the historical evidence confirms.

We cannot experiment on Aristotle, to be sure, but in his physics (innocent of modern chemistry, of course) the primary category of being consists of substances that indeed exhibit many of the properties I have just mentioned. Yet first, not all Greeks accepted such a picture. For the atomists in particular, atoms and the void alone exist, the atoms quality-less, indivisible and invisible entities, separated by the void that ensures both plurality (it separates the atoms) and move-ment (the atoms are in constant motion: on their own they move in no particular direction).[27]

Moreover, secondly, the Chinese generally viewed the world very differently, as can be illustrated by the misunderstandings that arose when the Jesuits first arrived there in the sixteenth century. They interpreted Chinese physics as a botched attempt at Aristotelian ele-ments which they themselves thought provided the correct solution to the constitution of things.[28] But the Chinese themselves spoke not of elements, but of phases. In their account of change they focussed on

[27] Yet Aristotelian qualitative four-element theory was opposed in antiquity not only by atomism, but also by various versions of scepticism, which withheld judgement on all matters to do with invisible entities and hidden causes. I shall return to that issue in chapter 5.

[28] Gernet 1985 provides a brilliant exposition of this encounter and of the misapprehensions on both sides.

*wu xing*, the five phases. These were named Fire, Earth, Metal, Water, Wood (I shall come back to the translation of those terms in a minute). They were linked in two cycles, one of mutual production, in which fire produced earth, which produces metal, which produces water, which produces wood, which produces fire to start the whole cycle off again. The other is a cycle of mutual conquest, in which fire conquers metal, which conquers wood, which conquers earth, which conquers water, which conquers fire, again to start the whole cycle over again.[29] To the Jesuits the issue seemed to be just one of four elements versus five. But that was to ignore how those phases were understood: it was completely to discount the possibility that the underlying ontology was quite different. The phases are not substances, but processes. In fact, a Chinese text makes the point quite explicit.[30] *Shui*, the term conventionally translated 'water', is 'soaking downwards'. *Huo*, 'fire', is 'flaming upwards'. So where the Jesuits—and before them Aristotle himself, come to that—would have seen *things*, the Chinese saw *events*. But if Aristotle would have been baffled, another Greek would have been more sympathetic. I am thinking of Heraclitus, for whom the cosmos as a whole is an ever-living fire, being kindled in measures and extinguished in measures.[31]

My thought experiment with the Chinese seeing fire, at least *huo*, as a process and the Jesuits seeing it as an elemental substance is of course reminiscent of Quine's explorations of the indeterminacy of translation and of his and Kuhn's of the notion of incommensurable conceptual systems or worlds.[32] Quine imagined an anthropologist in the field who in the presence of a rabbit hears an indigenous informant say 'gavagai'. But, Quine's point was, the anthropologist

---

[29] *Huainanzi* ch. 4, from the second century BCE, provides a clear statement of these two cycles: see Major 1993: 186ff.

[30] *Hong Fan* ('Great Plan') in the *Shang Shu*, or book of *Documents*: Karlgren 1950: 28 and 30 (cf. Lloyd and Sivin 2002: 259f.). The date of this text is variously estimated as between the mid-fourth and the early third century BCE.

[31] Heraclitus Fr. 30. Theophrastus was to criticize Aristotle's treatment of fire as an element, pointing out that it cannot exist without a substrate (fuel) and suggesting that it is always in a process of coming-to-be and is a kind of movement, and so should not be treated as a primary element (*On Fire* III).

[32] Quine 1960: 51ff., 71ff., on 'gavagai'. Both Kuhn 1970 and Feyerabend 1975 made much use of the notion that different cosmologies may be incommensurable. There is, to be sure, no neutral, theory-free vantage point from which evaluation can be made. But that does not mean that all discussion of alternative views is impossible. I shall be returning to this issue in chapter 5.

is not entitled to suppose that 'gavagai' means rabbit, rather than, say, 'mere stages or brief temporal segments' of rabbits, the appearance of the rabbit, the way it runs off into the bush, or, alternatively again, not just a single rabbit but the whole of rabbithood in the world (gavagai might be a mass noun, like gold or water in English[33]).

On the point that translation is always provisional one may agree, and as that proceeds, basic conceptual assumptions may need substantial revision. Again, Kuhn was clearly right to insist on the radical shifts in meaning that have taken place in the major paradigm changes that have punctuated the history of Western science. But it would be absurd to conclude that the chief actors in question were never in a position to understand one another at all. Copernicus certainly thought he understood Ptolemy, as did Galileo and Aristotle. Einstein has been thought not to have fully understood Niels Bohr, but that has not prevented commentators from making the comparison between them as if both were intelligible enough. Incommensurability should not be taken (as it sometimes has been) to imply the impossibility of any mutual understanding: that would be to set extravagant criteria for successful mutual comprehension in the first place. That can never be perfect. But that does not mean that all efforts at understanding are systematically thwarted.[34] We do not need to talk about communication in a foreign language to make the point, since even in English and among native English speakers comprehension is never exactly perfect, if you comprehend me. Anthropologists regularly remind us of the hazards of translation—as indeed do ancient philosophers. But I have yet to hear of an anthropologist who returns from the field announcing that she could understand *nothing* about the people she was studying. The point that some

---

[33] Quine's discussion of the ontogenesis of reference, in particular his account of mass nouns (Quine 1960: ch. 3), followed the standard practice among most linguists and philosophers of language of taking English as the prime target of analysis. Yet some features of English do not map well onto other languages, as consideration of the sense and reference of such Chinese terms as *shui* ('water') or *huo* ('fire') (discussed in my text) or others such as *jin* ('gold', 'metal') would have shown, although that should not mislead us into falling into the trap of supposing that thought is simply determined by language (cf. below, ch. 3).

[34] Warnings against reducing others' conceptual systems to one's own are salutary, but that should not lead to the conclusion that others' ideas are inevitably beyond our comprehension. The key points, to which I shall be returning in subsequent chapters, are that our conceptual system is no monolith and our basic concepts are indeed subject to revision in the light of what we learn from others.

preliminary understanding, however subject to revision, can be achieved is essential, and I shall be returning to it in later chapters.

Thus far I have suggested not just that we should not set inordinately high, impossible standards for mutual understanding, but also that indeed history provides good examples where quite marked differences are found in what is assumed about the nature of reality, including about what the world is made of. Alternative ontologies in that sense are there in the historical record. Rather than reject them out of hand as not making sense when they do not conform to our own assumptions, we should make the most of the opportunity they present to see things differently, indeed to entertain alternatives, challenging and difficult as that generally is.[35]

But now a further difficulty arises, over and above the problem of coming to terms with alternative ontologies. Some, such as Tooby and Cosmides and their colleagues,[36] have focussed on human evolution and insist that as the human beings we are, we all share not just our genes but also our basic cognitive capacities. The argument is that they were formed during the long centuries of human existence as hunter-gatherers in the Pleistocene and therefore common to the ancestors we all share. There are still plenty of disputed questions about how humans evolved from earlier hominids, including whether that was a once-for-all development. But strong as the arguments are not just for our shared biology, but also for the interactions of biology and culture, recently emphasized by Nettle among others,[37] the problem I have with some evolutionary theses is that wherever we may have started, we as humans have now ended up with an altogether greater diversity of solutions to how to live—not merely how to survive, but the best social arrangements for life—than is catered for simply by an appeal to our original shared Pleistocene ancestry. Culture may and indeed must reflect biology through and through: point taken. Sahlins (2008) even

---

[35] Thus for Quine's indeterminacy of translation, generally viewed as a problem, we might substitute a principle of the open-endedness and revisability of interpretation—an opportunity. Cf. Ingold's plea for entertaining alternatives, Ingold 2008, cf. 2000.

[36] Tooby and Cosmides 1992. Cf. Boyd and Richerson 2005.

[37] Nettle 2009 has recently underlined the differences between 'evoked' and 'transmitted' culture. Runciman 2009 in turn provides a sophisticated triadic analysis of the interplay of 'evoked' (biological), 'acquired' (cultural), and 'imposed' (social) factors in the evolution of culture. Cf. also Enfield and Levinson 2006, and Levinson and Jaisson 2006.

puts it that 'culture is the human nature'.[38] But then our notion of biology has to take into account the complexities of our cultures.

So that takes me back to anthropology, and to the alternative ontologies that Viveiros de Castro and Descola have proposed. Let me proceed outwards from the comparatively uncontroversial to the less so.

Discounting minor differences between individuals (some of us are tone deaf, others not, some are 'colour-blind' as we say—though that is an oversimplification[39]—and others not) we may say that there are considerable commonalities in the perceptual apparatus of all human beings. That applies to the ancient Greeks and Chinese as well as to the Achuar and the Araweté (the two peoples that have been the subject of detailed ethnographic studies by Descola and Viveiros de Castro respectively). Yet we know that other animals perceive things differently. Bats evidently navigate by sonar, so their sense of space is mediated by sound, not sight.[40] But as regards many other animals (including jaguars), no one (I suppose) is in any position to give an account of their perceptual world, let alone their inner experience, that is anything other than a projection of their own ideas. I observe that jaguars drink manioc beer in some parts of Amazonia, maize beer in others.

But from the point of view of my discussion here, we can finesse the problem of the jaguars' own experience and concentrate just on how that has been imagined, talked about, and used. Where other human beings are concerned, we can, thanks to the anthropologists, make a start at describing their lived experiences, their worlds in that sense, so different from the one we usually take for granted, and yet not totally beyond our reach. We can see that to think of jaguars as forming societies like those of humans says something not just

---

[38] As Geertz 1973: 49 put it, there is no such thing as a human nature independent of culture.

[39] See Lloyd 2007: ch. 1: 19.

[40] The question 'What is it like to be a bat?' was the subject of an influential paper by Thomas Nagel (originally 1974, cf. 1986), but that did not pay much attention to the detailed experimental evidence that was already becoming available on their navigational techniques and skills (cf. Griffin 1974). The 'what is it like' question has been intriguingly pursued in a wide range of contexts (e.g. Leach 1972: 11, what is it like to be a rat, and Agamben 2004: 39ff. on the non-communicating worlds of flies and spiders), especially, but not exclusively, in connection with the development of consciousness, as in Humphrey 2011, cf. Dupré 2002. Meanwhile, empirical studies of such phenomena as synaesthesia and blindsight further challenge traditional assumptions that each mode of perception has its specific sense-data, cf. Weiskrantz (1986) 2009, Humphrey 2006: 65ff.

about a putative jaguar ontology, but also about us humans. It is certainly to adopt a very different attitude from the one we normally hold to the question of how humans relate to other beings, for it suggests that we are not the only creatures to have well-ordered relations with conspecifics, to behave appropriately, to conduct ourselves with seemliness, decorum, and respect. It is this opportunity to stretch our imaginations and rethink some of our basic assumptions that is opened up by the anthropological reports, just as I would hope it can also be opened up by exploring ancient Greek and Chinese thought. Descola's animism and Viveiros de Castro's very different perspectivism both see humans as living in a world surrounded by creatures that are endowed, like us, with selves and personhood. Descola's naturalism, by contrast, cuts us off from other animals, with nothing but stuff, what we are made of, to share. What is at stake here is—among other things—how we should live. Ontology here, as often elsewhere, implicates morality.[41]

Yet a second, sceptical, challenge now has to be met. Do we not know enough about animals to unmask those speculations about the animal world as a whole? Did I not point out that some species do (but by implication others do not) have complex social organizations? To say that all animals have culture is to make a similarly erroneous generalization as the denial that any do, though in both cases the issue revolves around what is to count as 'culture'. That point has some force, as it seems to me. But the correct response is not to resolve the problem by legislating how 'culture' must be understood.[42] What we need is not definitional fiat, but an understanding of the complexities of the question. We should acknowledge the limitations of our present grasp of animal behaviour and psychology (the subject of so much ongoing research after all), and that means recognizing that some of our usual categories can be seriously misleading, not just the outworn dichotomy of nature versus culture, but also when we slip into generalizing about animals despite their incredible heterogeneity, or when we persist in assuming a Scala Naturae in which everything

---

[41] An appreciation of ontological pluralism may, at the very least, serve as an antidote to an unthinking preoccupation with the anthropocentric values of 'materialist progress'.

[42] The confusions surrounding attempts at the definition of culture were well emphasized long ago by Kroeber and Kluckhohn 1952; cf. also Wagner 1975 and Kuper 1999.

else is ranked by its proximity to us at the top of the pile, though again the question has to be raised who *we* are.

It is not just a matter of acknowledging that we still have a long way to go in understanding other animals in all their diversity: we don't know enough about humans either, about the different ways humans have endeavoured to make sense of their environment and to organize their lives and their relations with friends, foes, and everything in between. We still have a lot to learn from anthropology and indeed from history, about how humans see themselves in relation to other humans and the environment, about their—our—imaginations, hopes, and fears, and after all about our self-deceptions, errors, and delusions. We are not infallible, no more were the ancient Greeks or Chinese, no more are the Achuar or the Araweté.

So I come back to the question I started with of where we stand, as the humans we are. Traditionally it seems that many societies have used gods and beasts to arrive at an answer, though as we have seen, the ways they have done so vary very considerably. However, we may remark a persistent tendency to construct a scala and to find some *essential* characteristic to secure human distinctiveness, Aristotle doing it with his notion of human rationality, Xunzi with morality, the Christians with the notion that we are created in the likeness of God and have immortal souls.[43] Yet while all those views focus on aspects of what Descola would call interiority, animism, in his schema, takes up a diametrically opposed position, with the claim that what separates humans and other beings is just their bodies, their coverings, their clothes. Yet that too represents, of course, a simplifying move with all the complexities now relegated to physicality. However, that recurrent urge to identify a single differentiating factor is repeatedly thwarted by the amazing proliferation of the *actual* answers that are given. The risks of schematization are obvious: maybe we should learn to live without the prop of simple answers.

---

[43] Aquinas, however, was one who stayed closer to Aristotle. In his *Summa Theologiae* 13: 96.2 he wrote: 'Man was master of other things to the measure that he was master of himself. He shares reason with the angels, sense-powers with other animals, natural vital powers with plants, and the body itself with all non-living things.' Aristotle has no angels (though he does have a plurality of Unmoved Movers responsible for the motions of the heavenly bodies), but otherwise his construction of a scala, for example in *On the Generation of Animals* II 3, 736b31ff., with living beings differing in 'honour' or 'worth', is quite similar.

The two contrast sets traditionally invoked both have their snags. If gods are used, the trouble is that—as has always been known or at least suspected—views about them differ. That is analogous to the point I made about using morality, in the manner of Xunzi, though the situation here is appreciably worse, since faith or the suspension of disbelief has to play such a large part if anyone is to claim that his or her own imaginings in this area *alone* are sound.[44] Moreover, the insistence that there is just one true religion destroys the solidarity that might otherwise be produced by having God create all human kind. If some are true believers, others infidels, some are saved, others damned, that tends to undermine any sense of the common humanity of humans as humans.

But if other animals are used, the first risk is that mirage I just referred to, of dealing with all those animals as, well, just animals, and the second is that they are often cited in furtherance of the idea of human dominion and of human uniqueness, and here what is threatened is any sense of solidarity between living creatures as the living creatures we all are—a solidarity that is not just a matter of biology, but also one of morality.[45]

But if both gods and beasts are difficult to think with, how do we use humans instead? In two ways, I suggest. One obvious recommendation, which does not need me to spell it out, is that in programmes of research in cognition, psychology, social behaviour, and evolution, studies of humans and other animals need to be more fully integrated, more cognizant of one another, in some instances less blatantly Eurocentric, than has sometimes been the case. That is, to avoid both the uniqueness claims, and the types of global generalizations that I have criticized when the two domains, humans and non-human animals, are treated as polar opposites.

But then the second tactic I would recommend is to turn the problem into an opportunity. We can and should use all the resources available to us, from anthropology and from history, and come to that

---

[44] Although Boyer 1994, 2001, identifies certain cognitive constraints to which (he claims) all religious beliefs are subject, the particular forms those beliefs take are, of course, generally considered by those who hold them to be fundamentally different, and it is all too rare for alternatives to be tolerated.

[45] Quite what that solidarity should consist in, and the question of whether all animals, or at least some of them, have rights, and if so what they are, are, of course, highly disputed issues. See especially Singer 1976, Sorabji 1993, Beardsmore 1996, Clark 1997, Osborne 2007: 162ff., 195ff.

also from literature and art, to investigate how humans have got on with the business of self-understanding. To investigate what it is to be human is, among other things, to investigate the human propensity to investigate—to speculate, to produce cosmologies. That may sound too barren and self-reflexive to be useful, but I would disagree. We cannot do without ontology, without values, without notions of agency. So we had better pay attention to the gamut of possibilities that our evidence throws up. This is not so much a matter of pursuing research in cognitive science as exploring the plasticity of the human imagination, the range of human creativity. We need both to learn from the errors that we may be liable to commit—and there are errors that we must be prepared to diagnose even though that has so often been done out of mere prejudice—and then also to explore the imaginative possibilities that have been opened up, difficult as that so often is. It is to the further exploration of these issues that I shall devote the studies in the chapters that follow.

# 2

# Error

THE last chapter mentioned, but set to one side, the problem of error—a crux for anthropologists, for philosophers, for historians of science, indeed for anyone in any society. Most people would agree that humans are fallible. But on what basis can a given individual claim that some other person's ideas or assumptions are radically mistaken, whether that person belongs to the individual's own group or to some quite different one? How is error to be diagnosed without prejudice? Even if we cannot draw up neat rules to settle the matter, what general principles can be cited to provide some approximate guidelines?

I agree with those who have pointed out that all descriptions, all judgements, are theory-laden, none is totally neutral. But whose theory, whose conceptual framework, is presupposed? That of the person making the judgement, clearly. But if that judgement concerns someone else's ideas, their theories, is that not bound to distort them? If that judgement diagnoses error, how can it claim to be unprejudiced—a point one can anticipate being made not just by post-modernists? We return to the difficulty that to apply our categories to others' ideas is liable to distort them. Yet our categories, our conceptual framework, are the only ones we have.

We may start by accepting that many diagnoses of error stem from or involve a failure of understanding. But we need to preserve our intuition that not all do. Sometimes at least we recognize our own mistakes when they are pointed out to us. That certainly applies to relatively straightforward judgements of 'fact', mathematical calculations, for example, or reports of simple matters that we have mis-perceived or misremembered, or even when it is pointed out to us, or we discover, that we have made a faulty inference or jumped, as we say, to an unwarranted conclusion (cf. Tversky and Kahneman 1974, Kahneman, Slovic and Tversky 1982). Time and again in the history of science errors have indeed been corrected on the basis of more careful observation, though sometimes underlying ideological assumptions also have to be revised. Aristotle's opinion that the human heart has three chambers would be an instance of the first, his idea that queen bees are kings one of the second.[1] Sometimes alternatives to crucial underlying assumptions were simply not contemplated (Euclid evidently did not entertain the possibility of the denial of the parallel postulate) or were dismissed too swiftly (Ptolemy was aware of the proposal of the axial rotation of the earth, but ruled it out on the basis of the lack of any observable effects on clouds or missiles). But in general it is rather when more complex judgements are at stake that we may choose a line of defence that suggests that our interlocutor has not fully understood our position—and that no mistake has been made—a line open, of course, to our interlocutors as well.

Applying that to the types of problems that we encounter both in ethnography and in history of science, we can see that as a first move in assessing whether our judgements are well founded, it is vital to consider whether the mistake does not lie with us rather than with the people on whom we pass judgement. That may sound obvious, but may be extremely complex. Are we, to begin with, dealing with opinions or statements that can be judged true or false, or rather with visions where the appropriate criteria of assessment are rather their appropriateness or inappropriateness, their felicity or lack of it? Have we sufficiently considered the points of view of the agents themselves? What do we know about those points of view? With matters that fall within the scope of the history of science as traditionally conceived we may feel reasonably confident that the evaluation of them as theories or opinions is possible. But with the complex ontologies reported by anthropologists that may no longer be straightforwardly so. Hegemonic naturalists have often been too quick to pass judgement in both types of case. Thus when the Jesuits

---

[1] While Aristotle makes clear his general assumption that males are stronger, more intelligent, and more courageous than females, he acknowledges exceptions, noting that female Laconian hounds are cleverer than males, and that female bears and leopards are braver than the males, *History of Animals* 608a27ff., 34. Moreover, as regards the reproduction of bees, having set out his views, he explicitly notes that the data have not been sufficiently ascertained, and if they are, then perception should be trusted rather than arguments, though arguments can also be trusted when they agree with the appearances (*On the Generation of Animals* 760b30ff.), a principle not always adhered to by his successors nor by his critics.

encountered the Chinese five phases in the sixteenth century (as I noted), they imagined that the Chinese mistake was to fail to count air as an element and erroneously to consider wood and metal as such, and indeed the recognition that the phases are not elements at all, but processes, met with considerable resistance in Western scholarship until quite recently.

The debates that raged in the 60s and 70s, both within anthropology and outside it, on the problems posed by so-called 'apparently irrational beliefs', offer an example where attempts to secure a viable interpretation often revolved around the question of whether the reported paradoxical statements were intended to be taken literally, or merely figuratively. But since those who made the statements rarely made any such distinction, its application to the problems they posed was bound to be conjectural. I shall be returning to the difficulties inherent in the literal/metaphorical dichotomy in chapter 4. What those debates did clearly illustrate, however, was the cardinal importance of locating such statements in the contexts of all the background information concerning the conceptual framework to which they belonged (including the criteria the agents themselves used in assessing them), though to be sure even when the fullest contextualization has been achieved, problems of understanding often remain.

That takes me to the fundamental problem of how we can begin to understand not just apparently exotic beliefs and practices, but any conceptual system other than our own. Again, reflection on our own subjective experience can help us to make a start. We distinguish between the mundane matters of fact that I mentioned, and the more complex sets of beliefs that describe attitudes, values, and the like. When we say that we believe in freedom of speech, that is shorthand for quite a constellation of ideas, the limits or boundaries of which may be indeterminate, not to say fuzzy. When we say we believe in animal rights, it may well not be immediately apparent to our partner in conversation what we believe them to include. Nor should we expect that any two people who express such a belief will necessarily have precisely the same range of application for it. When it comes to someone saying that they believe that God the Father, God the Son, and God the Holy Ghost are three Persons but one God, explaining what that article of faith amounts to will be an even more convoluted matter. Of course a preliminary gloss might be that they believe what the Church lays down to be believed: but that leaves the question of what beliefs those are still to be clarified. I shall be coming back to some of the paradoxes of religious belief later.

The first point to be emphasized, then, is that what we call conceptual systems are highly complex bundles of ideas and beliefs, the precise implications of which may be unclear even to those who recognize them as their own. Indeed they may well not have *precise* implications as such at all. Yet that has not stopped them from being judged and evaluated. Theological Councils devoted many weeks and months of strenuous debate to decide what were, and what were not, correct interpretations of the Trinity. On those decisions depended such matters of consequence as whether a person was to be excommunicated, whether their souls were destined for heaven or hell. Tolerance was often in short supply—as it still often is.[2]

Such episodes highlight two points. Whatever we may feel about the difficulty of diagnosing error in complexes of belief, such errors have repeatedly been identified and anathematized in the past. But secondly, such episodes illustrate the prejudices that may be in play. It is the very fact that others' belief systems and ontologies have been so often condemned without a scrap of hesitation that underlines the importance both of intellectual caution and of moral empathy. No doubt the reasons one theological interpretation was preferred to another were always complex. But on occasions at least it seems that the participants were less concerned to make every effort to see the issue from their opponents' point of view than to make sure that those opponents were defeated. Battles over theological correctness were generally battles for authority and control.

The history of Western thought is punctuated with notorious examples of clashes between systems of beliefs, theological, political, philosophical, and scientific. I mentioned the application of Kuhn's notion of incommensurability to the paradigm shifts, as he called them, in the history of science, as between Copernicus and Ptolemy, for example, or between Galileo and Aristotle, or Einstein and Niels Bohr. Kuhn's focus was on how key terms changed their meaning,

---

[2] To diagnose others with whom one disagreed as mad was a tactic already much used in Greco-Roman antiquity, and that tactic acquired an altogether new dimension when the treatment of those so diagnosed involved incarceration in institutions devoted to the purpose. On the first point, see Lloyd 2003, on the second Foucault 1965.

planet, force, mass, weight, for example. But Galileo's confrontations with the Church serve to illustrate how a diagnosis of error, in that instance, masked a radical divergence in what the argument was about.[3] What was at stake on one side was the solution to an astronomical problem, including, to be sure, views on the correct methods for arriving at one. But the prime issue, on the other side, was the authority of the Church and its interpretation of Holy Scripture. In the 1616 interrogation, Cardinal Bellarmine imagined that if it were conceded that Galileo was right, that authority would be fatally undermined. At that point the technical arguments of celestial and terrestrial physics took second place—for Bellarmine—to what was thought to be essential not just for the good order of society, but for the well-being of the faithful and the salvation of their souls. In Galileo's condemnation in 1632 error meant disobedience to the Church for the Pope and his representatives, where for Galileo it was a question of whether indeed the earth moved: 'e pur si muove'.

Nor has it just been in the West that debates on what might otherwise be considered purely scientific issues have been influenced by political or religious factors. When in the eleventh century the Chinese polymath Shen Gua proposed a number of reforms to astronomical theory and to the regulation of the calendar, his views were blocked and defeated by bureaucrats in the Imperial Astronomical Bureau in the name of maintaining traditional practice. He was targeted by a rival politician who resented his administrative abilities and potential influence at court. Shen Gua ended up being impeached and removed from the capital (cf. Sivin 1995: III).

Two principles of good methodology may be suggested that help avoid prejudice in the matter of diagnosing error, though both need supplementing. The first is, as we said, that we should, so far as possible, recover the participants' own points of view before we impose our own, keeping to the actors' categories, as the anthropologists say, rather than using our, observers', ones, conscious that they may be inappropriate or anachronistic or both. That faces the problem that I mentioned before, namely that of accessing those actors' categories in the first place, where radically different ontologies may be involved and where what is at stake is not a mere fact of the matter,

---

[3] For a recent discussion of the issues at stake in the affair of Galileo's condemnation, see McMullin 2009.

but a vision or a way of life. Nevertheless, as a recommendation to open our minds to contemplate alternatives and be prepared to see that our own initial viewpoint is not the only possible one, the principle clearly captures an important ideal.

Then the second methodological recommendation that is worth bearing in mind, though it too does not go far enough, is what has been called the principle of charity in interpretation. That has been invoked in rather different ways by Davidson and by Quine (cf. Delpla 2001). Sometimes the principle just relates to the interpretation of logical connectives. More often it is extended to apply also to statements of beliefs, where in one version the aim is to have others' statements come out true, so far as possible, that is to say in *our* terms. Again, the evident problem that remains is whether those terms of ours are adequate to the task. The obvious danger is that of presupposing the correctness or at least the adequacy of those terms, and of the criteria they presuppose, in carrying out the evaluation.

A third, more important point where the human and humane dimensions of our intellectual stance enter in is that our own conceptual schemata must be flexible enough to permit revision. That will certainly not be the case if they are treated as monoliths, blocks of ideas that comprise a would-be comprehensive system. It is at this point that the further principle, that all interpretation is provisional, becomes important. That includes on matters that fall within the domain of science, though, to be sure, the extent to which conclusions must be seen as provisional varies with the subject in question. Some of the findings of modern science are more robust than others, though none is, in principle, unimpeachable.[4]

Provisionality and revisability are well illustrated when several contrasting accounts of the 'same' phenomena are possible, when those accounts indeed should not be seen as alternatives, but as complementary. In Lloyd (2007) I developed two lines of argument on this issue, what I called the 'multidimensionality' of the phenomena, and the plurality of possible styles of inquiry. To illustrate multidimensionality[5] let me take first spatial cognition, where,

[4]  It is as well to bear in mind this point about revisability, even though we naturally appeal to the most recent findings of our, modern, science to decide many questions.

[5]  As my examples of colour perception and spatial recognition will show, my notion of multidimensionality has nothing to do with cosmological theories of alternative or plural universes, sometimes discussed under the heading of the multiverse, nor with possible

following Levinson (2003), we may distinguish three different frames of reference, an absolute, an intrinsic, and a relative one. In an absolute frame of reference, cardinal points (for instance) give a fixed coordinate system. In an intrinsic one, front and back and right and left are judged according to the dominant or natural character-istics of the objects in question: we speak of the front of a car as where the headlights are, the front of the television set as the side that faces us, and in both cases the right-hand side is judged from the front. In a relative frame, by contrast, right and left vary with the individual reference point: our right is the television's left when we face it. These three evidently have different implications, but it is not the case that we should say that one is correct and the other two mistaken.

Similarly, with the much-discussed case of colour perception, where classifications in terms of hue, in terms of saturation, and in terms of luminosity opt for different criteria, but again do not, or rather should not, force a choice. They certainly present divergent accounts, but we should not say that one and only one provides the basis for a single correct classification.

The second idea, of styles of inquiry, is one that develops points from Hacking (1992) and Crombie (1994), and carries a similar plu-ralist message.[6] Among the different styles they distinguished are the taxonomic, the hypothetico-deductive, the use of statistical analysis, and the laboratory-based style of modern science. With regard to a certain set of problems, the findings of these different styles will differ, but once again they are complementary and it is not a question of having, or even being able, to decide between them. Moreover, 'styles' in my usage covers not just the methods, but also the aims of an investigation and even the subject-matter that it purports to deal

worlds as that notion has been brought to bear by David Lewis and others on the problems of modal logic.

[6] In his recent book, 2009: 17ff., Hacking surveys the varying *fortuna* of the term 'styles', from Crombie's 'styles of scientific thinking' to his own 'styles of scientific reasoning' and his—reluctant—reversion to Crombie's expression. As I explained in my 2007: 6 n. 3, I use the term 'styles of inquiry' not just (as they do) in relation to modes of investigation (such as the genetic, or the axiomatic-deductive styles), but also to draw attention to the possible divergent foci of inquiry within a given domain, as when different models of taxonomy, using different criteria, are in play in zoological classification, or different notions of causation are assumed in cosmology. I agree with some of Hacking's reservations about the potential looseness of the notion of style, but consider it a useful analytic tool that can be brought to bear, judiciously, to capture elements of the pluralism of ontologies. See below ch. 5. For a recent critique of 'styles', see Kusch 2010.

with. In many such cases, as also when multidimensionality is in evidence, the main error we can diagnose is the error of believing that we are faced with mutually exclusive and exhaustive alternatives between which a choice has to be made.

How far do such principles apply to the very notion of ontologies I raised in chapter 1? When one of Descola's 'analogists' is faced with one of his 'animists' or 'totemists' (if we can imagine such a situation), will they just talk past one another? They evidently will not just talk past Descola himself, since he is equipped with a hermeneutic schema with which to locate the fundamental assumptions in each of the four ontologies he identifies. Even if each of these four incorporates radically different understandings of such concepts as self, body, agency, nature, culture, those differences can be analysed, as Descola demonstrates by doing so so successfully. Yet as he also recognizes, the problems of interpretation are acute, with naturalists, perhaps, particularly prone to be dismissive of other ontological regimes. Some commentators would insist that, given that there is no neutral vocabulary in which to parse them, different ontologies imply mutually unintelligible worlds. Yet that seems too strong and open to the objection that the very diagnosis of divergent ontologies presupposes a level of understanding at least so far as that diagnosis is concerned. To comprehend a radically different ontology does not mean reducing it to our own (whatever that is), provided, as I have been urging, we allow the revisability of our own assumptions and the possible multidimensionality of what is there to be understood.

A comparison with a famous anecdote in the *Zhuangzi* compilation (4th to second century BCE) (17.87–91)[7] is suggestive, even though at one point the story depends on a play on words. The philosopher Zhuang Zhou (that is, Zhuangzi himself) is strolling with his friend Hui Shi along a weir on the River Hao, and when they see minnows swimming 'free and easy', Zhuangzi says that is how fish are happy. To that Hui Shi says to Zhuangzi that he is no fish, so how/whence can he know the fish are happy? Zhuangzi replies by saying that Hui Shi is not him, so how/whence does Hui Shi know that he, Zhuangzi, does not know the fish are happy? Hui Shi thinks he has won when he concedes that not being Zhuangzi, he cannot know about him, but by

---

[7] I cite *Zhuangzi* according to the Harvard Yenching edition, Supplement volume 20 (Beijing 1947).

parity of reasoning, Zhuangzi, not being a fish, cannot know the fish are happy. But it is Zhuangzi who wins, since he points out that in asking how/whence he *knew* about the fish, Hui Shi assumed that he did. Graham (1989: 8of.) commenting on the passage says it does not just depend on the ambiguity in the expression I have translated 'how/whence', the Chinese *an*, 'from where', 'on what basis',[8] when another more usual expression to ask 'how' would be *he yi*, 'by what means'. For Zhuangzi, all knowing is relative to a viewpoint, so there can be no answer to a demand for its source other than one that acknowledges such a viewpoint.

This is not exactly the perspectivism in play in modern ethnography, though it shares with that idea that the perceiving agent must be taken into account when judging a perception. Further, one moral that the story suggests is that even the expression of doubt about knowledge presupposes that what it would be to have that knowledge is intelligible. It is crucial, here as before, to unpack what 'ontologies' comprise and how they are to be assessed, a particularly difficult task when, as often, they are matters not of beliefs or theories, but rather of practices and of ways of being in the world. While we all communicate with one another ideas concerning particular objects and events, and those ideas are subject to correction, ontologies are, in principle, comprehensive accounts of whatever there is. As such, they are clearly not subject to straightforward verification or falsification. They are metaphysical, though that should not be taken to imply that they are meaningless (even though that is an inference that the Positivists drew). But if diagnosing simple error is not appropriate where a total ontology is concerned, plenty of particular statements made from within an ontology may be incorrect, and recognized as such by the individuals who make them. The interpreter can thus begin to grasp the senses and references even of problematic terms, where the interpreter's own initial expectations differ and will have to be modified as interpretation proceeds.[9] That is of course not to deny that much remains uncertain. As I said, what precise or even imprecise commitments are made is often beyond reach.

---

[8] So on that reading the answer to the question 'from where did Zhuangzi know that the fish are happy?' is given by the place, along the weir of the River Hao.

[9] I shall be returning to problems to do with sense and reference in chapter 5, where I shall introduce the notion of semantic stretch as an important element in communication that allows for the open-endedness of interpretation.

But does that mean that 'anything goes'? That would be an extravagant conclusion to draw. Just as we have every reason to be cautious in considering some set of beliefs, thinking that we fully comprehend them, so there is every reason to be cautious in endorsing any metaphysical system, the implications of which may be very unclear.[10] That applies to any particular such system, and a fortiori to any suggestion that we should sign up to them all. While suspension of overall judgement, when dealing with whole ontologies, is normally the most appropriate posture to adopt, certainly initially, that does not mean that all critical evaluation is rendered null and void, and that no judgement is ever possible. Indeed some judgement, some assessment, is inevitable, given that, as we said, no description is or can be entirely value- or theory-free, though the descriptions and judgements are— as I have also insisted—always subject to revision.[11] What we need are commentaries on ontologies, one might say, and not just verdicts, certainly not ones that mistakenly assume that the ontology we happen to start with will serve as a universal yard-stick.

Thus far an eirenic stance can be justified. But two challenges must now be met, the first from the side of morality, the second from that of logic. The value of studying alternative systems of belief lies first in that it helps us to see the limitations of our own initial assumptions, both about what there is to understand and how to understand it. Those systems of belief are not just matters of intellectual cognition, but also of ways of living, of values and attitudes to other persons and beings in the world around us. That in turn means that we can learn how others organize their lives, what they consider necessary for well-being and happiness. But while our increased understanding in such matters is always worthwhile, that does not mean that we can and should adopt all the evaluations, beliefs, and practices of the people from whom we try to learn. First, as regards beliefs, we may note that mistakes in attributing agency may be particularly liable to occur, given that the demand for an explanation, for example of some

---

[10] At that point some of the doubts of the Positivists are well grounded.

[11] Even within a single shared ontology, the issue may well resist simple resolution. Already in Greek antiquity the debate between atomism and continuum theory (both substance-oriented ontologies, though otherwise so divergent) was not one that could be settled by simple appeal to empirical evidence or to rational arguments or even to a combination of both. What was at stake was, among other things, the acceptability of providentiality (cf. Sedley 2007).

misfortune, is most readily met by the identification of some intentional being, human or non-human—when the explanation then takes the form of apportioning blame—a trait that is widespread in every type of society, our own, no doubt, included. Then, as regards practices, we have every reason not to leap to some verdict that some practice that is reported is senseless or inhuman or immoral: but that cannot mean that, once we have considered the matter in all its complexity, we condone any and every practice. Judgements in moral matters are no doubt the most difficult ones to make. But they cannot be ducked or avoided. Tolerance is badly needed: but it has its limitations when confronted by the intolerant. Rather we should recognize that to learn how to make better informed moral judgements is one of the possible pay-offs of the study of other people's beliefs and ways of life.

But I have yet to consider the domain in which error seems most conspicuously diagnosable and is indeed frequently diagnosed. This is where mistakes in logic are made, particularly when inconsistency is suspected. However, this too is a topic on which caution needs to be exercised. Where we have two well-formed formulae forming a contradictory pair, then by the Principle of Excluded Middle it is indeed the case that one must be true, the other false. The Principle of Non-Contradiction states that at least one of a pair of contrary statements must be false, but that leaves open the possibility that neither is true.

The problem here is that well-formed formulae are not the common currency in which complex ontologies or even values are discussed. Indeed well-formed formulae are not common in ordinary conversation in general, where our assertions leave much unsaid even when our intentions are clear. To be sure, ontological discussions often include simple statements to which those two Principles look as if they can and should be applied—and indeed that has often happened, generally with drastically distorting effects. But as I said before, the precise commitments of ontologically complex statements are hard to unravel, and indeed *precise* commitments may not be in question at all.

But is that not—a logician will protest—just a recipe for fudge and mystification? To be sure fudge is common, though we should recognize that ambiguity has its uses when it allows an interlocutor to explore multiple avenues of interpretation, when it is an opening gambit in an exchange that may then take different directions.[12] Nevertheless, in general clarity is a virtue in communication, and shortcomings in that regard may be met by legitimate demands, precisely, for clarification.

But it is one thing to be unnecessarily obscure, it is another deliberately to set out to mystify, as in many cases where a person or group is in business manipulating another for their own ends. That too is a difficult judgement to make, especially when it is outsiders who make it. But that is not to say that we cannot ever identify and condemn ideological interference and naked exploitation when we see them, guarding, always, as far as possible, against our own prejudices and assumptions.

Mystification is often the effect, even if it may not have been the intention, in many of the paradoxes and counter-intuitive beliefs that are common in religious dogma and have even been seen (by Boyer 1994, 2001) to be a characteristic of religious discourse in general. So the question here is whether it is appropriate to resist the application of those two logical Principles that I have identified. The faithful will be faced with a dilemma. If we take an article of Christian belief, that God is Three Persons, but also essentially One, then to allow the application of the Principle of Non-Contradiction is to have to accept that one or other statement must be false. If, on the other hand, the faithful insist that the Principle of Non-Contradiction has to be suspended, does that not threaten to make the statements meaningless? The point has been repeatedly debated in Christian theology, with some insisting that the suspension of the Principle is too high a price to pay, while others have willingly paid it and face the consequence that religious discourse is not subject to the same rules that apply to well-formed formulae.[13]

The more charitable line of interpretation would have it that the pair of statements posing the apparent contradiction should be understood as inviting the listener to explore how it may be the case that we

---

[12] The diagnosis of Chinese as an inherently ambiguous language is a typical example of Eurocentric prejudice, among other reasons because it underestimates the careful and deliberate use of open-ended expressions in communicative exchanges. Ancient Chinese communicators were no less capable of saying exactly what they wanted to say than we are.

[13] This was the point at issue between Leibniz and Honoré Fabri, in the seventeenth century, as Dascal 2006: 237ff. has shown. However, in chapter 4 I shall draw attention to the abuses of the appeal to the Principle of Non-Contradiction when the non-propositional features of communication are ignored.

can believe that God is indeed One and yet also Three, in virtue, say, of different attributes associated with the Three Persons of the Trinity. That is, of course, one path that has been taken in theological exegesis when attempts at reconciliation are made to save the doctrine from the charge of apparent incoherence, although some moves seem ad hoc or gratuitous. Elements of fudge and mystification generally remain: indeed one powerful trend in religious discussion is precisely to *insist* that the Trinity is a mystery, Three Persons but nevertheless One Person. On occasion, compromise was not sought, but strenuously rejected in celebration of the awe inspired by such counterintuitive claims. But if the mysterious is going to be comprehended only by the few, when does the mysterious become the strictly unintelligible? To admit to bafflement in matters to do with the divine shows laudable humility: but that view does not combine well with any desire to persuade others that your own faith is the one true one.

The diagnosis of strictly logical error, my argument is, is only of limited use. Identifying or suspecting unclarity is a more frequent option. But diagnosing that, let alone conscious or unconscious mystification, can only proceed after the resources of charity have been exhausted and the fullest exploration of possible senses and viewpoints has been undertaken. We all have experience of considerations of power and authority muddying the issue of what people are allowed to believe is true or false. But we should not leap to the conclusion that that is the case before the evidence that it is becomes irresistible—not that we have any algorithm for determining when that is the case.

But where, I must now ask, do the complex ontologies reported in the ethnographic literature come in my taxonomy of possible candidates for error? Given the nature of the discourse in which those ideas are expressed, to which I shall be devoting chapter 4, judgement often has to be suspended. When faced with multidimensionality and plural styles of inquiry we can and should treat the accounts in question as complementary, not alternatives. Yet mystification cannot of course be ruled out, not least because the members of the societies reported on themselves on occasion suspect it, just as they have no difficulty in entertaining the possibility that their informers are lying.[14] Doubts

---

[14] Every community and society, our own included, has, to be sure, its more and its less credulous members, constrained in different ways by the practical or other difficulties of checking stories, opinions, interpretations, and theories, and of resisting pressures to

about what some particular shaman claims are frequently expressed (Shirokogoroff 1935: 332ff., 389ff., Lévi-Strauss 1958/1968: 175ff.), even if the institution of shamanism itself is not challenged.[15] Contesting the institution itself would, of course, involve not just questioning existing power relations, but also rejecting the model of intentional agency that shamanism deploys and on which it depends. That would not mean rejecting agency as such, but resisting an appeal to it as a virtually omnipresent mode of causation. I shall be attempting a discussion of the circumstances in which analogous challenges could be and were mounted in ancient Greece and China in my next chapter. Meanwhile, to note a point of similarity between ancient and modern societies, individual Greek and Chinese persuaders, like individual shamans, were often greeted with disbelief and on occasion their motives called into question. Some disputatious Chinese thinkers were criticized for their useless suggestions,[16] while some Greek ones were castigated, in some quarters at least, for their immorality.[17] Yet in the pluralist situation of both ancient societies, both types of rebuke, both the charge of futility and that of serving to undermine good order, were levelled by other thinkers against those whose ideas they disagreed with. But of course the fact that diagnoses of error have often been tendentious does not mean that that is always the case—as I have been insisting all along.

At the same time, the problems with alternative ontologies are not matters that can generally, or even ever, be resolved by the application of elementary rules of logic, though this does not mean that we need

conform, a point that is only worth mentioning in the light of residual naïve assumptions of comprehensive 'primitive' gullibility.

[15] I shall be mentioning briefly, in my next chapter, the phenomenon of religious conversion, sometimes effected under duress, but sometimes a voluntary act. In the latter case especially it illustrates how dissatisfaction with an existing religious regime may lead to its wholesale abandonment.

[16] This was the criticism levelled at Hui Shi, for example, in *Zhuangzi* 33: 69–87, where the reaction to the 'cartloads' of his writings was: 'what a pity his talents were wasted'. Cf. Graham 1989: 77f. But plenty of Chinese thinkers were condemned by others on the grounds that their teaching led to disorder and confusion (*luan*) in the state: see Lloyd 2010a.

[17] Plato's main complaint against those he branded as 'sophists' in such dialogues as the *Protagoras*, *Gorgias*, and *Euthydemus* is that they were indifferent to the morality of the positions they discussed and failed to educate those they taught. But the recurrent charge, in Aristophanes (*Clouds* 139ff.) among others, namely that the sophists taught how to make the weaker or worse case seem the stronger and better, goes further, since it implies their immorality.

to change those rules and postulate alternative logics.[18] Of course, the Achuar and the Araweté themselves are not in a position to consider the possibility that the Principles of Excluded Middle and Non-Contradiction should not be applied to their statements. Those Principles are not part of their intellectual toolkit nor of the rhetorical techniques of persuasion they deploy. But more often those statements present, to the outsider, the challenges to interpretation that I have spoken of. If we have some modest success in our efforts at understanding what they have to tell us, we are likely to find that we have to revise our own starting assumptions, not our ideas about logic, but on the rather more important matter of those about the range of possible human experience of the world.

---

[18] I argued against this move in Lloyd 2004, ch. 4.

# 3

## Ancient understandings reassessed and the consequences for ontologies

ANCIENT science, as I am prepared to call it, provides an excellent opportunity to test whether or how far we are there dealing with radically discordant conceptual systems and the circumstances in which they may be challenged and revised. The documentation available to us is extensive and it offers the occasion to study the substantial changes that took place in some societies both in the understanding of the phenomena investigated, and in the methods used to investigate them. I have already used Greek and Chinese examples to illustrate the ontological presuppositions underlying notions of what it is to be human. I shall now attempt to go into the issues in greater detail. First, however, it is important to review the past history of the subject and to examine critically some of the myths that have surrounded the question of the origins of science.

Historians of early science concentrated, until comparatively recently, on Greek achievements. Greek science was where Western science all began. Archimedes, Ptolemy, Galen were the great heroes of the invention of rationality itself. True, the scientific revolution had to question their ideas, and the ancient Greeks were often represented as a positive hindrance to progress in the battles between the ancients and moderns that then took place (not that Aristotle and Galen themselves should be blamed for the way their ideas were used, especially when they were misrepresented as having produced closed, dogmatic, definitive systems).

But now all that has changed—thanks to radical shifts both in the historiography of the subject and in the philosophy of science. First, historical: the study of the achievements of ancient Mesopotamian,

Egyptian, Indian, and Chinese science has been opened up, thanks both to the publication of new corpora of texts and to the pioneering interpretative analyses of a handful of remarkable scholars. Joseph Needham, of course, put ancient Chinese science, and more particularly Chinese technology, on the map, pointing out that the Chinese were responsible for most of the inventions that were identified as key to the breakthrough to modern science, the printing press, gunpowder, and the compass among them.[1] Although Otto Neugebauer was far narrower in his interests, he was one of the first to draw attention to the achievements of ancient Near Eastern mathematical astronomy.[2]

Secondly, developments in the philosophy of science have profoundly affected our understanding. Gone is the old positivism according to which scientific theories were supposed to be tested against what were presumed to be theory-free observation statements. The defining characteristics of science itself have been problematized. So far as antiquity goes, many have questioned whether we can properly talk of 'science' as such at all. The ancients certainly did not have a word for it. Obviously there were no ancient laboratories. There was rather little experimentation, even. Yet other styles of inquiry, important still in modern times, were practised, classification based on systematic observation, deductive reasoning, the genetic style in which phenomena are explained in terms of their origins. The results of ancient investigations can often be said, with the benefits of hindsight, to be flawed, but we cannot judge science by its results. Who knows what features of twenty-first-century science will need to be revised before the century, or even the decade, is out? Rather, as a first approximation, we should focus on aims, and there we can certainly track how ancient investigators, like modern ones, strove to observe, predict, and explain the phenomena in the world around them, in short to understand them.

So if both the historical and the philosophical situations have changed dramatically in the last few decades, one could say that the time is ripe to attempt a reassessment. While most of the hard historical work has to be done, and is being done, in relation to specific

---

[1] The first volume of *Science and Civilisation in China* appeared in 1954. Twenty-four volumes have been published to date and some four or five are still to come.
[2] Neugebauer's masterpiece, *A History of Ancient Mathematical Astronomy*, appeared in 1975.

problems in specific subject-areas, we also need a synoptic view of what we can learn from ancient science as a whole, not a Grand Theory in the Grand Old Manner, but a general stocktaking of where we now are thanks to those important recent developments. In particular, we should make the most of the light that ancient science can throw on the strategic question of how ontologies developed, competed with one another, and changed. A close analysis of the similarities and the differences between different ancient traditions illuminates issues that are still relevant today, concerning the range of science and the tensions between tradition and innovation that it exhibits. My remarks on the ancient world in this chapter will be directed at Greece, China, and Mesopotamia especially—the relevant Egyptian data are in shorter supply and those from India are of very insecure date. In each case I shall attempt a brief survey of some of the main features of our current understanding—with special attention, where Greece and China are concerned, to the issues of the ontological assumptions that were made, the modifications they underwent, and how we should evaluate them.

The earliest sustained investigations of the world around us for which we have detailed evidence are those that have come to light, in recent years, from ancient Mesopotamia. Cuneiform texts in great abundance, which were discovered way back in the nineteenth century, have been made available in critical editions not just by Neugebauer, but also by Pingree, Hunger, Sachs, Parpola, Reiner, and others, and crucial interpretative studies by Swerdlow, Rochberg, Robson, and Brown have advanced our understanding of their importance. Neugebauer himself showed no interest at all in, indeed nothing but contempt for, astrology, so although his work on mathematical astronomy marked a significant breakthrough, the picture it created was quite lopsided. He tackled questions to do with the range of observations made and the analytic tools used to interpret them, but on one crucial major issue he simply missed out.

This was on the question of the overall strategic programme of the Mesopotamian scribes (they were known as *ṭupšarru*). Their chief original interest, as we can see from the classic omen text the *Enūma Anu Enlil* (hereafter *EAE*: it was put together some time between 1500 and 1200 BCE, although it incorporates even earlier material), was in the interpretation of signs or portents. What did the omens in the heavens signify? The *EAE* contains a wealth of examples where some celestial event is correlated with a terrestrial outcome. These predictions were usually expressed in the form of conditionals, if so and so (the sign), then so and so (the outcome, called the verdict). A couple of examples will do: 'If Jupiter approaches the Crook [a constellation], the harvest of Akkad will prosper' (Reiner and Pingree 1981: 41), or 'if Venus stands in the horn of the Moon, the king's land will revolt against him' (Reiner and Pingree 1998: 43).

But then a change occurs (as Brown 2000 and Rochberg 2004 especially have pointed out, though they differ somewhat on some of the details[3]). Sometime before the middle of the first millennium BCE what came to be predicted is not *just* what is going to happen to some kingdom or king on earth, but also certain celestial phenomena themselves. The *ṭupšarru* came to be in a position to say, for example, when a planet would become visible again after a period of invisibility, or when an eclipse would occur (lunar eclipses, at least, solar ones are more difficult because of their different visibility from different locations on earth). They had discovered regularities in the heavenly phenomena that enabled them to make certain predictions with a good deal of assurance.[4] Note that what had happened was *not* that the overall cognitive aim had changed, from prediction (for instance) to (say) explanation. Rather, within the category of prediction, what could be predicted changed and expanded.

We are used to distinguishing between astronomy and astrology, but that is, of course, an anachronistic distinction so far as Mesopotamia goes. It is far better to talk generally of the 'study of the heavens', leaving open the question of what that comprises (I shall come back to disciplinary boundaries from time to time later). The *ṭupšarru* were into prediction full stop. They (usually) expressed themselves confidently over the entire range of their interpretations—but (and here's the point) they came to be aware of cycles of celestial phenomena that were, in fact, altogether more determinate than most of the correlations they dealt with on earth.

[3] They disagree, for instance, on when confident estimates of the synodic phenomena were possible, and on when the concept of synodic period as such was available.

[4] Much remained, to be sure, beyond their reach and puzzlingly they still contemplate 'impossible' events, such as had been included in the canonical text *Enūma Anu Enlil*, which refers to eclipses happening not just at opposition but on other days of the month.

They did not, of course, *say* they discovered 'astronomy'. They did not have terms to demarcate regular celestial phenomena from other *EAE*-sanctioned correlations. In particular, they did not have a concept with which to pick out natural phenomena, as such, as natural. In any case, the heavenly bodies are gods, and wilful ones at that. The, or rather an, explicit concept of nature was not to be forged until much later (by the Greeks): that will be a central theme for us later in this chapter. But the *ṭupšarru* were in a position to give confident predictions of a range of celestial phenomena—and *to get it right*. Of course they did not always do so, and in some cases *whether* they had got it right was debatable. But in other cases it wasn't. The eclipse occurred or it didn't: there was not much doubt as to the *fact*, even though controversy could rage over *why* it did or *why* it didn't. To be sure, some of their other predictions also turned out right, about events on earth, revolutions, the death of kings, or the birth of princes, and those that turned out wrong could usually be explained away. The difference was that, where planetary visibilities, for instance, were concerned, there was more confidence and less need for explaining away.

Yet this new ability changed nothing so far as the Babylonian conception of the heavenly bodies as gods was concerned. The whole heavens continued to be thought of as divine. The names of gods and goddesses, as Brown (2000: ch. 2) has shown, were used of constellations, of individual stars, and of planets, to the point where it is sometimes difficult to be sure whether a reference to Šamaš (say) is one to the deity or to the sun, or even, according to Brown, to the planet Saturn, or whether the name Enzu refers to a star (the Goat Star), to a constellation (Lyra), or to a planet (Venus). The new confidence in prediction was grafted onto traditional beliefs and did not replace them.

The crucial tools the Mesopotamian *ṭupšarru* used were mathematical ones, including, eventually, the construction of periodic tables, linear and zig-zag functions, that express, mathematically, the cycles of heavenly movements. Yet it is quite clear that they were also interested in mathematical relations for their own sake, as the list of so-called Pythagorean triplets in one cuneiform tablet (Plimpton 322) reveals.[5] This gives me my opening to my next ancient civilization, China.

---

[5] The list of triplets that Plimpton 322 sets out, all exemplifying the rule that $a^2 + b^2 = c^2$, shows that these mathematicians were well aware of the underlying relations. Many of the

Again, let me start with mathematics, where the two major classic texts, dating from 100 years either side of the millennium, have been well known for some time. These are the *Zhoubi suanjing* (Cullen 1996) and the *Jiuzhang suanshu* (the *Nine Chapters on Mathematical Procedures*) (Chemla and Guo 2004). But to those we can now add an earlier, less comprehensive and less systematic text, the *Suanshushu* (Cullen 2004), recently excavated from a tomb that was sealed in 186 BCE. Thanks to the existence of that text, we are in a better position to trace the early development of Chinese mathematics than we are for Greek mathematics before Euclid—not that what Chinese mathematics (*shu shu*, or *suan shu*) covered was identical with what Greek *mathēmatikē* did, although there is enough overlap to do a comparison between them. Two judgements about Chinese mathematics that used to be common can now be seen to be badly mistaken; certainly they need substantial qualification.[6] The first is that their interests were merely practical, and the second that though they were able arithmeticians, they were hopeless geometers. A single example will serve to show the inadequacy of both those ideas, namely their investigations of the circle-circumference ratio, in other words what we call $\pi$.

For practical purposes a value of 3 or 3 1/7 is perfectly adequate, and such values were indeed often used. But the commentary tradition on the *Nine Chapters* engages in the calculation of the area of inscribed regular polygons with 192 sides, and even 3,072-sided ones are contemplated (the greater the number of sides, the closer the approximation to the circle itself, of course): by Zhao Youqin's day, in the thirteenth century, we are up to 16,384-sided polygons (Volkov 1997).

Throughout the *Nine Chapters* the problems are expressed in concrete terms. The text deals with the construction of city-walls, trenches, moats, and canals, with the fair distribution of taxes across

mathematical tablets are school exercises (see, for example, Robson 2009), but nonetheless interesting for that, since they throw light on mathematical education and the formation of a cadre of numerate scholars. The impact of the formation of an elite on the further development of an inquiry may be either positive (in fostering collaborative research) or negative (when the elite constitutes an exclusive cadre entry to which is restricted on one criterion or another).

[6] They continued to influence Needham's interpretation of Chinese mathematics, both in the third volume of *Science and Civilisation in China* and in later publications.

different counties, the conversion of different quantities of grains of different types, and so on—all issues that needed to be dealt with in the increasingly complex societies of China in the so-called Warring States period (480 or 475 BCE to 221 BCE), let alone once China was unified by the first emperor, Qin Shi Huang Di, in 221 BCE. But those practical concerns turned out to be a trap for many unwary modern commentators. The interest in the original text is not just in practicalities, but in the exact solution to the mathematical problems.[7] At one point (ch. 5: section 6), for instance, we are told that the number of workmen needed to dig a trench of particular dimensions is 7 427/ 3064th labourers. The answer is not rounded to the nearest whole number.

But the Chinese also put their mathematics to work in both astronomy and harmonics. As in Mesopotamia there was intense interest in the study of the heavens, including not just the interpretation of omens (*tianwen*, that is, the 'patterns of the heavens'), but also the calculation of calendrical and eclipse cycles (called *lifa*). Note that this gives us ancient Chinese actors' categories for the two main subdivisions of the study of the heavens. Some have wanted to see *tianwen* as the equivalent to our 'astrology', *lifa* as our 'astronomy', but in neither case is there anything like an exact fit. *Tianwen* includes descriptive work, as well as predictions, and *lifa* was limited to arithmetical calculations of a variety of kinds. In antiquity it never included the construction of geometrical models to explain planetary motions, though that cannot be said to be a necessary condition for their study.

Here too, as in the ancient Near East, the subject was of high political importance and that ensured state support, though it also led to some interference. Huang Yilong (2001) has recently brought to light that some eclipse records were fabricated, and others omitted from the record—in both cases for ideological reasons. Although the astronomers understood that eclipses follow regular patterns, they were still generally considered ill omens. So fabricated records cluster in the reigns of 'bad' rulers, while omissions occur particularly when 'good' ones were on the throne. This provides an excellent example where with due caution *we* can bring to bear our knowledge (via

---

[7] In an interesting recent discussion, Chemla 2003 has suggested that the Chinese concern was more with generality than with abstraction, though there is certainly some overlap between them.

retrospective reconstruction of the eclipses that actually occurred) to check ancient records. As I have explained before, we generally have every reason to hesitate about the application of our, later, knowledge to the interpretation of ancient studies, for that so often leads to anachronism. But the evaluation of ancient eclipse reports is something of an exception. Even in this case, of course, we have to be sure that we are indeed dealing with an eclipse report, not a record of some other phenomenon. But on many occasions there is little or no doubt on that score. When we do that checking, we discover both how accurate in general those records were, and also that they were not immune to political manipulation.

As for the support given by the authorities, in Han times, the emperor Wu Di founded the Astronomical Bureau, which came to be staffed by literally scores of officials (we know their ranks and rates of pay). They were put to work not just to ensure that the calendar was in good order, but also to keep the emperor informed of impending eclipses and the like. In the latter case it was important that the emperor was not caught out by an eclipse that had not been foretold, for that would surely be interpreted as a sign that his Mandate from heaven was on the decline. The assumption was that the emperor was the key link between the microcosm of the state and the macrocosm of the heavens, and it was his responsibility and privilege to keep the two in harmony with one another. As long as that was the case, his Mandate was unassailable. But if there were untoward events in the heavens or on earth, his position could be vulnerable, and his enemies could use such phenomena to suggest he was finished.

In relation to the work on the calendar it is worth remarking that the Chinese (unlike Aristotle, though like Plato) did not generally assume that the heavens were completely unchanging. So far as those two Greeks go, Aristotle thought he had evidence—from the Babylonians indeed—that no substantial changes had ever taken place in the positions of the fixed stars (this was before the discovery of the precession of the equinoxes by Hipparchus in the second century BCE), while for Plato the changeability of the heavens followed from the fact that they were visible, physical objects. The more empirical approach led to Aristotle jumping to a premature conclusion: Plato's a priori handling of the problem allowed him to be open to the suggestion that the appearance of a lack of change could be misleading. In China, especially when one dynasty succeeded another, the new one imposed its own calendar partly for political purposes—to show who was in control—but partly on the assumption that the calendar needed to be kept up to date—as indeed it did. One of the principal modern authorities, Jean-Claude Martzloff (2009), speaks of the Chinese belief in 'celestial indeterminism', which had, as its consequence, a sense of the need of 'perpetual reform', which indeed he documents over some 1,600 years.

The second area of application of mathematics that I mentioned was harmonics, and again the convergence of literary and archaeological evidence gives us a far firmer basis for interpretation than used to be the case. The Chinese did not just study the theoretical problems posed by the twelve-tone scale: they constructed sets of chime bells that delivered that scale.[8] The mathematical difficulties they encountered were predictable. In the interpretation of scales you either have to allow approximations to keep the analysis to moderately small numbers or else you have to tolerate complex ratios such as that between 32,768 and 59,049, that is 2 to the power of 15 to 3 to the power of 10, or 2/3 to the power of 5 times 4/3 to the power of 5. Interestingly we find *both* approaches used in one of our main literary sources, the *Huainanzi*, compiled under the auspices of Liu An around 136 BCE. Yet where Greek harmonic theorists fought bitterly over whether an arithmetic or a geometric analysis was to be given, and again over whether perception or alternatively reason was to be the criterion,[9] there is no sense of any conflict or tension in the *Huainanzi* on the correct methodology.

One important underlying idea was that the heavens and the earth should be in harmony with one another. The *qi* (breath/energy) of the pitch-pipes should resonate with the *qi* of the seasons. We are dealing with one example of the widespread interest in the correspondences between things. But as I noted in chapter 1, the underlying ontology was not one of substances or elements,[10] but of processes. The five phases (fire, earth, metal, water, wood) are in constant interaction. They were brought into correspondence with large numbers of other

---

[8] The set of some sixty or so two-tone chime bells excavated from the tomb of the Marquis of Yi, buried in 433 BCE, provides the most remarkable proof of the application of harmonic theory in practical music-making: see Chen 1987.

[9] The controversy has been brilliantly analysed by Barker, e.g. 2007.

[10] Needham was well aware that the five phases are in constant interaction, but persisted in using the term 'elements' of them.

'pentads', groups of five items, ranging across heavenly bodies, planets, colours, musical notes (as in the pentatonic scale) all the way to rulers, ministries, and emotions. On the one hand, this is an ontology that pays particular attention to the analogies between things, indeed a classic case of what Descola called 'dizzying analogism' (2005: ch. 9) running riot. On the other hand, the things in question are not viewed as static independent or self-subsisting substances, unchanging in themselves, but rather as interacting shifting items, each seen not in isolation but in its manifold relations with other items.

The fundamental opposition between *yin* and *yang* illustrates the interactiveness very clearly. *Yin* gives way to *yang*, but at the height of *yang*'s influence, *yin* begins to re-emerge. Conversely, when *yin* is at its peak, there are already glimmerings of the re-emergence of *yang*. As a general rule, no item, no person for instance, is exclusively *yin* or exclusively *yang*. An old man, for example, may be *yang* as male as opposed to female, but *yin* as old contrasted with young (cf. Sivin in Lloyd and Sivin 2002: 198–9).

This sustained Chinese interest in correspondences led among other things to the notion I mentioned that pitch-pipes, buried in the earth, will reflect the seasons of the year. That turns out to be an example where what seems fanciful to us, seemed fanciful also to later Chinese investigators. As Huang Yilong and Chang Chih-Ch'eng showed (1996), the practice of what was called 'watching for the ethers' (that is, *qi*) was discontinued when the idea on which it was based was debunked as indeed quite unfounded. But we also have to note that those interests in correspondences prompted the Chinese eventually to make such discoveries as that of the directionality of the compass. The Greeks knew of the 'magnetic' stone, but never noticed the connection with the polar axis.

Programmes of research in China were usually geared to matters considered of importance to the state, to rulers, or after the unification to the emperor, and they were often carried out by officials working in state Bureaux. Yet thanks in part to the principle that everything, the heavens included, is subject to change, ideas and theories were acknowledged to be revisable and were revised. Meanwhile, on moral and political issues there was anything but general agreement about such questions as whether humans are intrinsically good, bad, or indifferent. This was the subject of a hard-hitting debate between Mencius, Gaozi, and Xunzi in the fourth and third centuries BCE (cf. Graham 1989: 117ff., Lloyd 1996a: 27ff., 77), where what was at stake was how much direction and control humans needed, issues of evident importance for those who saw one of their chief roles as that of advisers to rulers on good government. I shall have more to say on all these matters in chapter 4.

Our understanding of what the Greeks attempted and achieved has undergone its sea changes too, partly thanks to technical work done on the Greek materials on their own (especially by combining the findings from the archaeological and epigraphic records with those from our literary sources), but also partly by bringing a more ecumenical, or at least comparative perspective to bear. I shall focus on just two important issues. The first is the definition and practice of axiomatic-deductive demonstration, where we can make some progress on the two questions of how and why this new style of inquiry (as we may call it) was developed. The second relates to the development of an explicit category of nature, the basis of a new ontology, indeed several, where the issues to be explored include the circumstances in which this occurred, and how deep-rooted the new conceptual framework became. Both those ideas have been hailed among the triumphs of Greek rationality: but as we shall see, in both cases that is an oversimplification.

Part at least of the history of the development of the notion of strict demonstration can be traced, but first we need to be clear on what that notion involved. Aristotle was responsible for its definition, although he allowed other, looser kinds of demonstration as well, and not just in rhetoric (Lloyd 1996b: ch. 1). But according to him, *strict* demonstration proceeds from self-evident premises (axioms, hypotheses, definitions) via valid deductive argument to incontrovertible conclusions. Other traditions of mathematics, the Chinese for instance, have their notions and practices of proving, in the sense of checking or verifying results, including showing, for example, that an algorithm is correct. But they never developed any such notion as the one that Aristotle defined and that Euclid exemplified in practice—and that thereafter became the goal of *one* tradition of Greek mathematics (though there are important exceptions, as for instance in Hero of Alexandria[11]).

---

[11] The recovery of the pluralism and heterogeneity of Greek mathematics has been an important theme in the work of Netz 1999, 2009, Cuomo 2001, and Tybjerg 2004, among others.

Given that the Chinese got on perfectly well *without* axiomatics, producing a wide range of results both in arithmetic and geometry (as I have indicated), the puzzle is not why the Chinese did *not* have axiomatic-deductive demonstration, so much as why the Greeks *did*.

We can supply part of the answer by considering the situation in which investigators worked in ancient Greece. Aristotle, like Plato before him, reacted very critically to the merely persuasive arguments (as they thought them) that were commonly used by the professional teachers they called 'sophists' as well as in political and legal contexts. An audience, Plato has Socrates say, can be persuaded of almost anything, true or false: what was needed was a way of guaranteeing the truth. Strict, axiomatic-deductive demonstration did precisely that. It was invented, one might say, as a means of producing such knock-down arguments.

This is another example of the well-known competitiveness that characterizes so much of Greek culture, the agonistic spirit to which Burckhardt (1898–1902) drew attention in so many areas of Greek life, athletics, musical competitions, competitions for the best tragedy, and so on and so on. In public debate in all sorts of contexts, legal, political, as well as 'intellectual', for example in philosophy and science and even in medicine, you demanded that your opponent gave an account (*logon didonai*), justifying his policies, his actions, his theories. The judges in the legal and political domains were the dicasts in the law-courts and the citizen-body collected in Assemblies or Councils (they were the same people[12]). The decision was taken by majority vote—a strange institution which we take for granted, sometimes not noticing that it can distract attention away from an effort to find a consensus.[13] In what I have called intellectual debates it was your peer group, or sometimes even a lay audience, who decided who had had the better of the argument.

Thus it was all about persuading the relevant constituency, but Plato and Aristotle demanded more—demonstration producing certainty. They represented that mode of argument as quite different

---

[12] The dicasts acted as both judge and jury, deciding points of law as well as guilt or innocence and passing sentence, and they could number in their hundreds, even up to the entire role of 6,000 jurors at Athens in the fifth century.

[13] In many egalitarian societies decisions are taken by a massed assembly, but this is generally by acclaim (cf. Detienne 2003). In the Greek law-courts the dicasts' vote was secret.

from (mere) persuasion. It was above the competition: yet a way of winning that competition. Demonstration was the most effective means of persuasion, however much it might present itself as *not* to do (merely) with convincing an audience. All of that is well known: but perhaps its significance, especially in a comparative perspective, has sometimes been missed.

Of course, Chinese philosophers and mathematicians also engaged in debate (though more often their criticisms were directed at extant texts rather than with living opponents who could answer back), but in the background is the assumption that the ultimate authority—even in what we would call scientific matters, in astronomy, for instance—lay with the emperor or his ministers, who did, in practice, intervene in some of the discussions that are recorded (cf. Cullen 2000, 2007). Moreover, as I have already said, their chief mode of *proving* in mathematics was via the checking of algorithms, not axiomatic-deductive, and the whole quest for incontrovertibility would have struck them as misplaced.

This takes me to the second more complex and trickier issue of nature, that is, the emergence, in fact the invention, of the explicit concept, where the emphasis must be on 'explicit'. Of course, the regularities the Mesopotamian *ṭupšarru* discovered in the heavens were regularities in what *we* call *natural* phenomena. An *implicit* notion of the regularities of (certain) phenomena is present not just in that work, but whenever a farmer sows seed to produce a harvest. But it is one thing to have such an implicit idea, another for it to be made explicit—where *natural* phenomena are identified as a particular set of items, contrasted with those that are *not* natural, whether because they are 'supernatural', or because they do not belong to nature, but, for example, to culture, which came to be construed, in that context, as the antonym of nature.

We can see the importance of this development by reflecting on the fact that initially, the Greeks themselves had no such explicit concept: you will not find it in Homer or Hesiod for instance. It was some of the early natural philosophers (*phusikoi* or *phusiologoi* as they were called), and some of the medical writers, who started talking about *phusis*, nature, as such. In Xenophanes' case, one context in which he may have done so (but the evidence is fragmentary, Fr. 32) was his insistence that Iris, the rainbow, is (just) a cloud, where it may be that we should understand that there is no need to think of it—as Homer

had done—as a messenger from the gods. Our sources for the Hippocratic writers are more extensive and unequivocal. In the treatise *On
the Sacred Disease* the author maintains that that disease (epilepsy,
we may say, according to his quite detailed description) has its nature
and its cause just like any other. Craftily he does not say that none of
them is divine: rather all are, but that is because nature itself is divine.
All have natural causes and so there is no need to introduce personal
gods as responsible or as agents who might be persuaded to effect a
cure. We saw in chapter 1 that Aristotle, for instance, still stayed with
the notion that humans are between gods and the other animals,
though his idea of the supreme deity is no Zeus but an Unmoved
Mover. But the writer of *On the Sacred Disease* makes nature itself
divine and so also all the natural objects it comprises, even diseases.

But as I argued when I first tackled this problem in 1989 (my
Spencer lecture published in Lloyd 1991: ch. 18), this was no mere
intellectual move, the result of some piece of abstract analysis. These
natural philosophers and medical writers used the concept of nature in
order to carve out a domain over which they were to be the acknowledged authorities. This was ultimately about the rivalry between
those competing for prestige, the traditionalists who took their stand
by what had always been assumed (the gods are at work everywhere)
and the naturalists who claimed that was nonsense, a category mistake, and that they could offer far better accounts of the phenomena
(whether they could is another matter). 'Nature' is the watchword
they invoked to contrast their way of understanding with that of their
opponents. We may here suggest a contrast between different modes
of trying to outdo your rivals. The Babylonian *ṭupšarru* and Chinese
astronomers often accused one another of being ignoramuses, of
getting their facts or their interpretations all wrong, but they did
not generally do so by way of an epistemological argument, challenging the very foundations of their opponents' approach. The notion
that your rivals were wrong not just because they held mistaken
views, but because their methodological assumptions were at fault,
is a distinctive feature of Greek polemic.

But this Greek rivalry was not just with the traditionalists, but with
each other. We have, in fact, an amazing proliferation of ontologies, of
accounts of what there is. The earliest natural philosophers, Thales
and Anaximander in the sixth century BCE, appear to have been
interested in the origins of things rather than in their physical

constitution. The first writer to have an explicit notion corresponding to 'elements' is Empedocles, whom I shall be considering in a moment. But if I am right about Thales, for instance, is interest in origins had something in common with even earlier accounts, such as Hesiod's *Theogony*, even though that spoke of the births of gods and goddesses, while Thales had everything originate from water, a substance that he almost certainly considered divine, but certainly did not think of as a person.

But from then on we get one type of account after another. Heraclitus flatly denied that the world-order had a beginning, saying in Fr. 30 that it is an 'ever-living fire, being kindled in measures and extinguished in measures'—an idea that, as I said before, might have made him more sympathetic to Chinese ideas than most Greeks would have been. Parmenides was an approximate contemporary of Heraclitus: we do not know who was the elder, nor can we be sure that either knew the other. But Parmenides adopted a view that went to the opposite extreme, in that he flatly denied both change and plurality. Being, for him, was unchanging and unitary. That is the subject of what he called the Way of Truth: the rest is just a Way of Seeming or Appearances, on which mortals wander, he says, deceived by their senses and their opinions. Parmenides was evidently prepared to say that everyone else was mistaken and to do so on the basis of what he took to be an irrefutable argument. Nothing can come to be from nothing. But nothing *comes to be* from what is already existent either. For that already *is* and so does not *come to be*. Yet from the outset it was obvious that Parmenides, the proposer of the argument, was himself subject to the world of appearances.

Later Presocratic philosophers did not accept Parmenides' conclusion, but did see the need to reinstate change, namely by postulating a plurality of some kind. The situation was thus very different from that which obtained in China, since there was no Chinese Parmenides who denied that change could occur. So the common-sense experience of coming to be and passing away was not threatened and needed no justification. In fifth-century Greece, Empedocles, as we saw in chapter 1, thought that all living things are akin, but physically they are constituted by earth, water, air, and fire—the four 'roots'—which are combined and separated by two cosmic forces he called Love and Strife. All six of these are divinities, though markedly different from the anthropomorphic deities, such as Aphrodite (for instance), who were the object of traditional worship.

Though both Plato and Aristotle take up Empedocles' four elements and adapt them to give their own accounts of the constitution of material objects, other Presocratic philosophers were far from following Empedocles' lead. Faced with Parmenides' denial that plurality and change are possible, the most economical response was that of the fifth-century atomists, Leucippus and Democritus, for whom atoms and the void alone are real. Everything else, the qualities hot and cold for example, exists by 'convention', *nomos*, alone, where the term also used for 'culture' is now applied to perceptible qualities in order to suggest that they only had a marginal or derivative claim to reality, one that depends on our, human, conventions. The void is that which separates the atoms and through which they move, thereby securing both plurality and change, but the atoms themselves are homogeneous, solid substances, differentiated by shape, size, and arrangement alone.

Plato and Aristotle both reverted to different versions of a qualitative account of matter, based on the four simple bodies, earth, water, air, and fire, adding that any cosmology also had to pay attention to the final causes of things. For both of them the principal ontological question that needed addressing was not the material constituents of physical objects. Rather for Plato, ultimate reality consists in invisible, intelligible Forms: the particulars we perceive in the world around us merely participate in, or imitate, them. While Aristotle contradicted that view in no uncertain terms (as we shall be mentioning in chapter 4), he agreed with Plato that any account of substances must include one of their essences or forms, even though for Aristotle these were aspects of those substances, not transcendent realities.

From the perspective of a comparative analysis of ontologies, this plethora of theories defies generalization. But six points are worth highlighting. The first observation is that each theorist seems set on outdoing predecessors and contemporaries alike, to produce an account of the world that would see off all rivals, combining substantive points with methodological and epistemological ones to achieve that end. This is in the competitive spirit I discussed in relation to the development of the notion of strict demonstration. Secondly, the claim is that what is offered is an account of reality that goes beyond the appearances: the truth of the philosopher's own system is contrasted with others' mere opinions. In those circumstances an appeal to what we can perceive or what is generally assumed has no standing. Thirdly, in most cases the account of what is is an account of *nature*. Even for Plato, the world of Forms is the true nature (*Timaeus* 52b, cf. *Parmenides* 132d). Fourthly, while in their debates competing theorists often seem to talk past one another—and Aristotle in particular can be especially narrow-minded and dismissive in his evaluation of his predecessors—it is nevertheless the case that in general rivals saw themselves as addressing the same problems, at least when those problems are stated in abstract, general terms, as the questions of what is and what can be known. In that sense they would not have accepted that they were fundamentally at cross purposes, namely because they were trying to understand different worlds, let alone inhabit them.

Fifthly, with rival views of nature also went rival views on one of its antonyms, *nomos*, covering laws, customs, and conventions. One group, who included Plato, sought to integrate human laws with divine ones, subordinating the former to the latter. A second, represented in texts in Sophocles, Herodotus, Thucydides, Xenophon, and Demosthenes among others, distinguished varying human laws from unwritten, universal ones, though these were always hard to specify (cf. Lloyd 2005: 120–1 for the details). But a third, more common position, associated, for example, with the views that Plato puts into the mouth of Callicles in the *Gorgias* and Thrasymachus in the *Republic*, accepted the great diversity of human laws and customs, and argued either that each community creates those it adopts to suit itself, or that those who were able to impose their views on the rest of society, if not indeed also other peoples, would inevitably do so, embodying the principle that might is right.

Sixthly and finally, beyond these disputes about morality and politics, there was a fundamental disagreement on the question of the nature of any proper account of the cosmos and its parts. Teleologists, such as Plato and Aristotle, insisted that such accounts should refer to final causes, the good that phenomena manifest, in Plato's case the intelligent, benevolent activity of the Divine Craftsman who does not create the world from nothing, but introduces order into chaotically moving matter, and in Aristotle's case the craftsmanship immanent in nature. For him an account of a part of the body, such as the hand, must pay attention not just to what it is made of, but the way it serves the good of the creatures that have hands, namely humans. But against this the anti-teleologists, the Atomists especially, excluded teleological explanations and appealed purely to the physical interactions of things. For them the infinite worlds they postulated do not result from design, but from the chance collisions of atoms. Given that the issue was whether or not the cosmos is the creation of some providential force, this is an excellent illustration of the point I made before (ch. 1), namely that ontology implicates morality.

We saw that Chinese thinkers differed on the inherent moral characteristics of human beings, and in chapter 1 I cited a text from the *Lüshi chunqiu* among the many that recognize the diversity of customs. Yet the ideal that the microcosm of human society should resonate with the macrocosm of the heavens is common ground to most of those who were in business, in China, offering advice on good government. So for the Chinese there is no radical split between two domains, one of nature, the other of culture, that have accordingly thereafter to be reintegrated with one another. True, for the Greeks, humans, as natural animals with customs and laws, straddled both nature and culture, but the introduction of that explicit dichotomy, where *nomos* in the sense of social institutions and morality belongs to humans alone, opened up a gap with fateful consequences. That divide lies at the origin of so much subsequent Western soul-searching, where nature and culture are treated not as parts of a seamless whole, nor as complementary aspects of a multidimensional human experience, so much as antonyms or polar opposites that must somehow be rendered compatible with one another.

At the same time all this abstract Greek speculation and debate must be seen against a background of a generally conservative system of traditional beliefs. Most ordinary Greeks, there is every reason to suppose, continued to believe what their forefathers had believed, not just that they could, up to a point, accept the evidence of their senses, but also that there were gods (usually thought of in human form) at work in the world, to whom sacrifices should be offered according to the practices that tradition laid down. We can illustrate this from the domain of medicine. Although the Hippocratics and later theorists launched a scathing attack on the idea that gods were personally responsible for diseases or could cure them, just those beliefs continued to flourish in the context of the practice of temple medicine, arguably the most successful, in the sense of most popular, mode of healing throughout pagan Greco-Roman antiquity and even for some time beyond (cf. Lloyd 2003). It was all very well for the writer of *On the Sacred Disease* to substitute his impersonal notion of the divine for one that invoked gods as wilful agents. But his own claims to understand the causes of epilepsy, let alone his assertion, at the end of the treatise, that most diseases are curable by adjusting regimen, involve large elements of wishful thinking.

Moreover, some deflationary remarks are in order regarding the two chief ideas I have been focussing on, the idea of axiomatic-deductive demonstration and the explicit concept of nature. They have regularly been hailed as major advances without which science could not progress. That they are remarkable, indeed idiosyncratic, ideas is clear. That science could not get on without them is given the lie by the achievements of the other two civilizations I have brought into the picture. But we should also reflect on the problematic character of both ideas.

As for axiomatics in the Greek sense, of securing primary premises that are themselves indemonstrable but self-evident—very different from Hilbert's axioms, of course, and the point is important—such primary premises were always more difficult to come by than the advocates of the method allowed. Mathematics provides the best examples, as for instance the equality axiom, take equals from equals and equals remain. Even in that sphere, Euclid's parallel postulate (whose status was already challenged in antiquity) is more problematic than Euclid himself appears to have imagined. But outside mathematics, when medical writers such as Galen, or philosophers such as Proclus, advocated proof *more geometrico* in physiology or theology, the premises tended to be either truisms (and vacuous) or controversial and indeed controverted. That search for incontrovertibility, the argument that would silence the opposition once for all, could be a distraction from growing the subject: demonstration can get in the way of heuristics, as examples from the history of mathematics, both in antiquity and down to the seventeenth century, show. Archimedes himself illustrates this dramatically when he demands that the results obtained by his new heuristic method, set out in his treatise of that name, have thereafter to be demonstrated by the standard techniques of *reductio*: you assume the contradictory of what you want to prove, and then show that that assumption leads to a self-contradiction, thereby establishing the point you set out to demonstrate.

Then as regards nature, quite what that is supposed to cover is—still—intensely controversial. We take for granted that the natural sciences have natures as their subject-matter. And yet a lot that is currently investigated in laboratory science does not exist 'naturally' in the sense of independently of the situation that brings it into being in the laboratory. That poses quite a problem for any naïve realist view of what science investigates, for the objects investigated are created by the procedures of investigation. Yet that should not lead the unwary to go to the opposite extreme and advocate an equally naïve relativist or instrumentalist theory, according to which what science studies is just what the scientists of the day decide is there to study. Then exactly what 'human nature' should be held to comprise is even more problematic, as is how culture, society, upbringing, and nurture impact on what is natural as in some sense what is given—given in our genes, ultimately, but only as a potentiality, whose actualization can take different forms (cf. Keller 2000). These are issues that are still far from resolved, despite all the efforts of social psychologists, ethologists, evolutionary theorists, and others to crack them.

The study of Mesopotamian, Chinese, and Greek investigations has thus recently revealed quite a lot of diversity, as well as points of shared interest, in the changing patterns of understanding of the world. Evidently each of these ancient societies developed its own science in response to its own needs and circumstances. Each had its own classification of the different important areas of investigation, where *our* categories of 'mathematics' and 'astronomy', let alone 'physics' and 'biology', are all more or less *in*appropriate. That is, (once again) not to endorse a blandly relativist or subjectivist stand—to the effect that each group is the sole arbiter of what science *is* and of how it is to be divided up. Rather, it is to adopt a pluralist position. All three ancient societies engaged in systematic attempts to achieve a better understanding of the world around them. On that score, in my view, all rate as science: it would be ridiculous to claim that just one tradition represents science in antiquity to the exclusion of the other two. Rather, we have to learn to adopt a more ecumenical, less parochially occidental view of the development of science itself. Better history of science, I would claim, contributes to better philosophy of science, as I shall be endeavouring to show in chapter 5.

There was, there is, no single path that the development of such systematic inquiry had to take, or should have taken, no royal road to science—by which I mean no unique route and no single goal.[14] The proper task of the historian of early investigations is to reveal that diversity and to ponder the reasons that contributed to how and why that occurred—a difficult task, in all conscience, where further pitfalls await the unwary, in the form of global hypotheses concerning linguistic, political, or ideological determinism. These are issues on which I shall have more to say later, but a couple of remarks are germane to our discussion in this chapter.

First, the very variety of the ideas and theories produced in all three ancient societies, and expressed in each case in the same natural language, gives the lie to any suggestion that they all, in each case, stem from certain characteristics of that natural language—let alone that they all stem from the mere fact of the presence of literacy. The point is obvious if we think of the enormous differences between what Democritus and Aristotle both had to say in Greek, or both the *Zhuangzi* and *Huainanzi* in ancient Chinese. Democritus held that there are infinite worlds separated by the void, while Aristotle was a continuum theorist and believed the cosmos is unique and eternal. The *Huainanzi* adopted the ontology based on five phases that I outlined in chapter 1. The *Zhuangzi* developed a radical sceptical philosophy of language. Starting from the differences in what people deem to be so and the shifts in the reference to which such claims relate, Zhuangzi concluded that asserting that something is the case, or indeed that it is not, is unjustified.[15]

Political factors and values too obviously have a great influence—as I have noted not least in relation to the level of support given to investigations and the creation—or the lack—of institutions in which they can be pursued. But again it would be extravagant to claim that politics or ideology determined *all* the inquiries that were carried out in all their complexity and heterogeneity. We need also to bear in mind that while politics undoubtedly influences intellectual life, there is also considerable feedback from science and philosophy on values

[14] It is tempting to see this in quasi-evolutionary terms: but if so, it is important to remember that biological evolution is not single-track either. For the analogies and differences between biological, cultural, and social evolution, see especially Runciman 2009.

[15] *Zhuangzi* 2. 27–32, cf. Graham 1989: 178ff. Although Zhuangzi disallows asserting that something is the case (*shi*) or is not (*fei*), he concedes that you have, in a sense, to rely on things being so (*yin shi*). I shall be returning to other aspects of the philosophy of language in *Zhuangzi* in chapter 4.

and ideology. In this area too we must pay due attention to pluralism and diversity, and I would be the first to underline the work that remains to be done before we are in a strong enough position to say precisely *why* these three ancient societies developed the particular modes of inquiry they did. But *that* they each opened up new avenues of understanding is the key point that confirms that very possibility, the existence and exploitation of *that* degree of plasticity in their traditional systems of belief.

So what can we learn from the comparative study of ancient science that can throw light on the problems to do with ontologies? Both the Chinese and, more especially, the Greek data suggest the difficulty of generalizing about the ontology of either ancient civilization. At the same time, and by the same token, more positively, they enable us to identify some of the conditions that prompt or enable the proliferation of ontologies, whether we are talking about the proposals put forward by articulate writers or the background of commonly shared assumptions. Let me elaborate both points briefly in conclusion.

In all three ancient civilizations we have considered that a highly literate elite existed, whose size at any given time cannot be quantified, but which was in each case significant. In China after the unification certainly, and in certain fields long before, there were elite cadres working in official capacities in state institutions with access to considerable archives accumulated over generations. The importance of such records for the type of astronomical work undertaken in all three civilizations is obvious. In Babylonia and China the running of considerable states was a further stimulus not just to develop bureaucratic structures, but also intellectual tools, for example in mathematics. In classical Greece the situation was very different. There was no empire to run, nor emperor to persuade, though there were of course autocrats to deal with, especially in the Hellenistic period, and eventually Roman emperors. In each of the many autonomous classical city-states the chief political decision-takers were a body of citizens, who also constituted the potential audiences for those who aimed to make a name for themselves for wisdom in any one of a number of fields of would-be expertise, Masters of Truth as they have been dubbed.[16] They had the leisure for

---

[16] The term was coined by Marcel Detienne 1967/1996 for particular use in relation to the intellectual leaders that preceded Parmenides, but it has a wider range of application

philosophical, cultural, and other agonistic pursuits, thanks of course to the surpluses produced by an economy based largely on slaves.

Such conditions evidently do not apply to non-literate societies. It is a mirage to suppose that such societies are incorrigibly static and homogeneous, though, as I have noted before, concrete evidence for diachronic change is often harder to come by. While, as is well documented, intellectual or religious leaders, including shamans for example, compete with one another for prestige, power and influence, in small-scale societies the state structures do not exist that enable the establishment of institutions geared to the accumulation and transmission of new knowledge. But as I argued in Lloyd (2009), some such institutions could have a quite ambivalent role, resisting innovation (when their elite members saw that as threatening their own position and authority) as often as positively encouraging it.

Then the second point worth noting is that even when the general conditions I have mentioned do exist, they do not always have the same consequences. The modes of investigation, scientific or philosophical, undertaken in the three literate ancient civilizations we have discussed differ in the ways we have seen. What stands out, in the Greek experience, is the strident adversariality among those Masters of Truth. But those rivalries were not just a stimulus to innovation, but also one reason why so many of the new ideas that were introduced gained little acceptance. No sooner had one bright idea been proposed by one Master of Truth than it would be contradicted by another with his own suggested solution to the problem.

Traditional ideas and beliefs could be and indeed, as we have noted, were exposed as unfounded also in China, but those tendencies were far more strongly developed in Greece, where we encounter radical subversion, by particular individuals, even of the practices of the city-state religion. In their place, ideas that do not just seem bizarre to us, but did so to the ancient Greeks themselves, were proposed in great profusion, with only a handful of occasions when their proposers got into trouble for doing so. When they did so, as in the case of

than that period, since individuals who claimed to be special possessors of the sole truth continued down to the end of pagan antiquity. I shall have more to say about the contexts in which they operated in the next two chapters.

Socrates,[17] they were prosecuted not by agents acting on behalf of the state, but by private individuals. That is an important contrast with the situation later in Europe, the one that faced Giordano Bruno or Galileo, although the positive analogies between Greece and the Renaissance included the proliferation of theories proposed by ambitious intellectuals competing for prestige.

The Greek speculations I have surveyed evidently purport to provide answers to the question of what there is. Ancient Greece would seem to be one of the moments in human history when such speculations were especially prolific and heterogeneous, though the majority of them had in common the claim that what they offered was an account of *nature*. That was the key to the rhetoric with which many new ideas were presented. Nature was contrasted (as we saw) on the one hand with the supernatural, the realm of what these new investigators generally dismissed as superstition or magic, and on the other with culture, *nomos*, the domain of human morality. Those Greek investigators shared with modern 'naturalists' as defined by Descola not just a notion of nature that unites all physical objects, but also a notion of culture or morality that is specifically human. The old contrast between humans and the other animals, that had been part of the triad, gods, humans, beasts, still generally survived in various forms.[18] Yet with some Greek naturalists, those contrasts were eroded, first, in that animals were seen to belong to nature just as much as humans do, and secondly, in that, as we saw in the Hippocratic *On the Sacred Disease*, it was nature that took over the role of what was divine.

Those Greek Masters of Truth were individuals who were not learning directly from others, but leaving convention behind to head off in new, often very surprising, even we may say counter-intuitive, directions. In the process they provided many illustrations of a capacity to innovate even in matters of deep-seated beliefs, including on what things are made of, or the self, or gods, or causation and

---

[17] Yet the charge that Socrates did not adhere to the city-state religion was pretty clearly trumped up. There is no suggestion otherwise in our sources that he refused to participate in normal religious activities, and he certainly thought highly of the Delphic oracle, which pronounced that no man was wiser than him.

[18] Yet the recurrent use, including in writers such as Aristotle, of animal characters as models for human ones—the lion is the paradigm of courage, the wild boar of stubbornness, the fox of cunning, and so on—still provided plenty of scope for the appreciation of parallelisms between the human and animal domains: cf. Lloyd 1983: 18ff.

responsibility. That was what they needed to do in order to make their reputations: so they had every reason to be innovative. Yet so often those new ideas had little or no general impact. It was all very well for Greek philosophers and scientists to propose new solutions to the problem of what there is, but that remained largely just a matter of their personal beliefs and those of a handful of their immediate followers, however much they might claim that they had been incontrovertibly demonstrated.

Descola's disinclination to denominate ancient Greece as a society with a naturalist ontology is understandable, but problematic. It is understandable, since the majority of the population (insofar as one can talk about them) remained unmoved by the theories of the elite that revolved around the notion of nature that they in a sense invented. But it is problematic for two reasons. First, as noted, it is certainly the case that the main components of naturalism and of the multiculturalism that forms its counterpart were already available in ancient Greece. Secondly, as regards the question of the degree of penetration of those concepts beyond the circles of the elite who proposed them, we have to register that similar reservations apply to much later periods of European thought down to the supposed triumphs of so-called modernity in the last couple of centuries. Those ongoing residues and complexities tend rather to underline the difficulties of arriving at fully satisfactory global characterizations of the underlying conceptual systems of whole collectivities, whether by using ontologies to do so, or indeed under any other generalizing rubric.

So one of the lessons we can derive from a study of the conditions of possibility of new ontologies in ancient civilizations is that they were generally the speculations of individuals—admittedly in distinctive political, economic, and social conditions—not matters that affected the beliefs of the society as a whole. The main exception to that remark relates not to philosophical debate, but to religious conversion, when mystery religions and especially Christianity eventually *did* impinge not just on the elite, but on considerable masses of the population. The reasons for this were no doubt complex and various, though in Greco-Roman antiquity superior technology or political power had nothing to do with it.[19] But it is obvious that while the

---

[19] There are evident differences between conversion under coercion, when one society or group imposes its religious beliefs and practices upon another, and voluntary

philosophers dispensed wisdom and freedom from anxiety, and if they spoke of immortality, supported that as far as possible with reason and argument, the advocates of the new religions offered salvation for eternity, often combining that with the threat of damnation for the unbeliever. Moreover, as we shall see in the next chapter, Christianity in particular backed up an insistence on faith and the suspension of disbelief with radical, counter-intuitive proposals on the nature of religious discourse and the sources of its legitimacy. Meanwhile, from the point of view of the elite, the conversion to Christianity sometimes appeared as a *reversion* to the traditional ideas that that elite had called into question. The triad of gods, humans, beasts notably re-emerges, now transformed, with the distinction between the last two drawn more sharply than ever, in an effort to allow for a new conception of the proximity of humans to God.[20]

conversion, when those who change their views and customs do so because they see some advantage to themselves in doing so—in terms of political position, or preferable morality, or a greater chance of happiness whether in this life or in the next. The possibility that the perception of those advantages may change is illustrated by the phenomenon of a reversion to an original faith. This has been studied by Vilaça 2010, 2011, for instance, in connection with the Wari', who switched to Christianity when they were persuaded it could yield such benefits as releasing them from inhibitions about eating certain animals as well as providing the wherewithal to fight disease. Yet when fatal diseases continued to occur, the Wari' reverted to their earlier belief that this was the result of sorcery. Moreover, a second related point was that Christianity was thought to block human reproduction insofar as it had the effect of making everyone consanguines, while for a human group to reproduce itself it needed affines, of course, even though affines were traditionally conceived as potential enemies.

[20] God became man indeed, though Jesus Christ was no ordinary man. The contrast between the Godhead and fallen man remains strongly marked, even if the message is that humans should strive for redemption and to be like God. Meanwhile, although the other animals too are God's creation, they, unlike humans, are not made in the image of God.

# 4

# Language and audiences

THE task this chapter sets itself is first to examine how far the content of opinions (from global cosmological theories to judgements on more mundane specific questions) may be influenced either by the communicative context in which they are expressed or by the status of the speech acts in which they are made. This will take me, secondly, to some observations on the much-debated issue of how language relates to the world. Can an examination of that question throw light on the problems of mutual understanding that I have been addressing in these studies?

We are all familiar with the phenomenon of speakers tailoring their messages to the audiences they address. How far is that relevant to the problem I have mentioned of understanding others' ideas, whether in the history of science or in ethnography? A favourite device (as I have remarked before) that was used to interpret, explain, or more often explain away, surprising, paradoxical, or totally counter-intuitive beliefs was to invoke the distinction between the literal and the metaphorical senses of terms, even though as an *explicit* concept that was often, indeed usually, not available to the participants in the original expressions of belief themselves. The study of hermeneutics owes much of its origin to attempts to interpret the paradoxes associated with Christianity, the Trinity, Transubstantiation, the Virgin Birth, the Assumption, the Ascension, the Death of God, and so on.

I shall first examine critically the application of that dichotomy. I shall then turn to a second favourite contrast, that between myths and rational accounts, a pair of concepts much used in different ways in the interpretation of the ethnographic as well as historical data. The third section of the chapter will consider what we know about the typical audiences faced by Chinese and Greek speakers and writers, where I shall follow up my earlier suggestions that the nature of those audiences does have important repercussions, in both ancient societies, not just on the ways in which ideas are presented, but also on their actual content. My claim is not that the character of the audience is always a crucial factor influencing the substance of what is presented, but the more modest one that it sometimes is, and notably in the case of some Greek and Chinese work. Turning back from that foray into ancient history to the general interpretative problems of understanding others' worlds and the question of the way language connects to reality, I shall conclude with some tentative suggestions for the analysis of the ontological claims involved, developing the idea I offer as an alternative to the literal/metaphorical dichotomy, namely that of semantic stretch.

Much ink has been spilt on the analysis of the distinction we still tend to take for granted, between the literal and the metaphorical. It is generally assumed that the literal is, in some sense, normal, the metaphorical deviant. But in what way it is deviant is far from clear. A number of paradoxical and counter-intuitive notions reported in the ethnographic literature were at the centre of debate on 'apparently irrational beliefs' in the 60s and 70s. How were reported Nuer statements that 'twins are birds' to be taken? Both sides to the dispute generally took it that one could meaningfully apply the literal/metaphorical dichotomy, both those who insisted that such statements were to be taken literally and those who argued rather that they should be understood metaphorically or figuratively.

Yet that pair could certainly not be said to be actors' categories. On the contrary, the dichotomy has a determinate, European, origin and history.[1] Aristotle, in fact, was principally responsible. Before him Plato had certainly frequently referred to images, likenesses, paradigms, often commenting critically on their potentially deceptive character. But Aristotle was the first to define what he called *metaphora* as a transfer or the 'application of a strange term', where he classified the four main kinds as (1) from genus to species, (2) from species to genus, (3) from species to species, and (4) by analogy, in Aristotle's sense of a four-term proportional analogy, as A is to B, C is

---

[1] I have devoted several studies to the analysis of this question, the first being in 1987. A recent monograph devoted to metaphor (Rolf 2005) distinguished no fewer than twenty-five different theories, while a study limited to metaphor in Greco-Roman antiquity had a bibliography running to thirty-nine pages (Lau 2006).

to D.[2] That analysis seems innocent enough, but in practice he, and others, frequently deployed—as many still do—the category of the metaphorical for polemical purposes, to downgrade the ideas and statements where he diagnosed a failure of literality.

Two examples from Aristotle himself will serve to make the point, in both of which the metaphorical is associated with the poetic. Empedocles' idea that the sea is the sweat of the earth will not do, Aristotle says (*Meteorology* 357a24ff.). It may be adequate for the purposes of poetry, for metaphor is poetic, but it is not adequate for understanding the nature [of the thing]. Similarly, when attacking Plato's theory of Forms (*Metaphysics* 991a20ff.) Aristotle says that to say that they [the Forms] are 'models and that other things share in them is to speak nonsense and to use poetic metaphors'. It is striking that, dealing with the key ontological views of his great predecessor and teacher, he dismissed them as based on a flawed use of language. Evidently, for Aristotle, poetry will not do for philosophy or science. I shall in due time give reasons for dissenting from that view.

The demand Aristotle himself makes is not just for the literal, that is, the 'strict' (*kurios*) use of terms, but also for univocity, for only if univocity is maintained across a stretch of reasoning will that reasoning be valid. His analysis thus connects to his theory of strict demonstration which I discussed in chapter 3. It is only if the terms in the premises are strictly univocal that the conclusion can be said to follow. Yet this is another instance where Aristotle's official theory is one thing, his actual practice another. His discussion of *metaphora* in the *Rhetoric* book III is itself full of what he himself has to admit are metaphors (cf. Lloyd 1996b: ch. 10).

At the risk of labouring the point, but to introduce a new element in the discussion, we may turn to the Chinese to show first, that not every culture, once it embarks on the analysis of the use of terms, will come up with the literal/metaphorical dichotomy, and secondly, that an alternative approach to the interpretation of others' communications is possible. Already long before the unification, the Chinese were interested in language use, with, in most cases, a very different set of preoccupations from those of the ancient Greeks. Thus what the Chinese called the rectification of names, *zhengming*, is a topic explored

---

[2] Aristotle, *Poetics* 1457b6ff.

in texts from the *Lun Yu* (*Analects*) attributed to Confucius onwards. Yet this turns out to be less to do with making sure words matched (in some sense) objects, than with securing the establishment and maintenance of proper social roles and statuses (cf. Gassmann 1988, Lackner 1993, Lewis 1999: 31ff.).

But more important for our purposes is the discussion of language use in *Zhuangzi*, ch. 27, which has indeed been thought—quite incorrectly in my view—to adumbrate a theory of metaphor. This text distinguishes three types of saying: *yu yan*, translated 'lodge sayings', *zhong yan*, 'weighty sayings', and *zhi yan*, 'spill-over sayings'. Several features of the interpretation of this text are problematic. 'Spill-over sayings' apparently refers to a kind of vessel that is designed to 'tip over and right itself when filled near the brim', which Graham (1989: 200ff.) glosses, rather optimistically, as referring to a 'fluid language which keeps its equilibrium through changing meanings and viewpoints'. 'Weighty sayings' are clearer in that these refer to 'what you say on your own authority'. Much in this text remains obscure, but the main point for my present discussion is clear. Although *yu yan* has often been compared with metaphor, and indeed the expression has sometimes been translated 'metaphor', the lodge sayings in question are where you 'borrow a standpoint outside to sort the matter out', that is, you put yourself in your interlocutor's position. Not only is this far removed from a notion of 'metaphor', but neither of the other two 'sayings' independently, nor both together, provides a category remotely resembling the 'literal'. Zhuangzi's suggestion is that you adopt your interlocutor's point of view. This, in the context, is to persuade him, but evidently the notion is relevant also to the problems of communicative exchanges more generally, where the fundamental point is that you should take into account the point of view of your partner in conversation. I shall have more to say about this idea of Zhuangzi's later.

But even though most of those whose statements were the subject of those debates in the 60s and 70s did not themselves have any actors' category that was the equivalent to that of the metaphorical, that does not by itself mean that that category is useless for the interpretation of those statements and the corresponding beliefs. Yet in practice the application of that category led to an unresolved dilemma. If for the Nuer twins are literally birds, you would expect them to have the characteristic features of birds, feathers, beaks, wings, and so on. But according to Evans-Pritchard (1956: 131, cf. Sperber 1985: 59), the Nuer denied that. But Evans-Pritchard also reported (1956: ibid.) that the Nuer insisted that twins are not just like birds, but *are* birds, rejecting thereby the figurative or metaphorical option. But instead of taking this—as others did—to confirm that the Nuer entertained a number of irrational beliefs, it should rather have been concluded that the literal/metaphorical dichotomy was not applicable.[3] We shall be considering later (ch. 5 p. 109) what might be taken to be an analogous paradox from Christianity, where for believers, in the Mass the wafer is not like the body of Christ (indeed it is not), but *is* the body of Christ.

The problem of finding an alternative framework for linguistic analysis is exemplified not just in ethnographic reports, but also and even more dramatically in another instance from theology, where we are dealing with a tradition of scholarly interpretation that stretches from ancient down to modern times, and the beliefs in question continue to be held today by members of our own society. From early Christianity I may take the examples of the death of Christ and the Resurrection. God cannot die, and yet the second person of the Trinity was crucified on the cross. That dilemma was famously greeted by Tertullian (around 200 CE) with a reaction that did nothing to mitigate, but positively gloried in, the contradiction. 'The son of God was crucified', he wrote, 'but that is no shame, since it is shameful' (*non pudet, quia pudendum est*). Again, 'the son of God is dead: this is to be believed, since it makes no sense' (*credibile est, quia ineptum*). Then again concerning the Resurrection: 'Having been buried, he rose again: this is certain because it is impossible' (*certum est, quia impossibile*) (*On the Flesh of Christ*, ch. 5). The Principle of Non-Contradiction, like the dichotomy between the literal and the metaphorical, generally forces a choice. One or other of the contradictory statements must be rejected. Either a term is to be taken literally or it is metaphorical. But Tertullian's response is to suspend the

---

[3] As it was, Evans-Pritchard resolved the issue, to his own satisfaction at least, by saying that 'in respect to God twins and birds have a similar character' (1956: 132). Whereas in many instances the ethnographers themselves admit to being baffled, it is foolish for the rest of us to engage in armchair speculation as to the meanings of the paradoxical expressions in question, even though the notion of semantic stretch that I introduce below would allow for more possibilities than the two alternatives, of the literal and the metaphorical, generally permit.

principle and turn that into a demonstration of the special nature of the Godhead. However, any such move requires, one would have said, some explanation, not the mere dogmatic assertion that God-talk is unique. Many later Christological disputes, notably the Arian controversy in the fourth century CE, turned on whether or not one should take literally the begetting of Jesus or his death.

So far as Tertullian himself goes, he was well aware of the mainstream of pagan Greek philosophy and of Aristotle's insistence on the Principle of Non-Contradiction. But that was no deterrent. The power of his message and that of other Christians came, to be sure, from another quarter, especially from the doctrines of the afterlife, the promise of heaven for the faithful, and the threat of hellfire for the evil. Those were considerations that proved to be more effective in gaining converts than any of the rational arguments of pagan philosophy. Arguments were no longer what counted: the truth had been revealed in a sacred text. The task of the faithful was not to question, but to believe. As Tertullian put it in *On Prescriptions against Heretics* ch. 7, we have no need of research after the gospel. 'When we believe, we desire to believe nothing more. For we believe this first, that there is nothing else that we should believe.'

Then a second dichotomy that has been brought to bear to help with the hermeneutic problems is that between myth and rational account. Once again, the Greeks invented the dichotomy, or at least one version of it. The situation is complicated by the fact that both of the terms usually treated as antonyms, namely *muthos* and *logos*, have a wide range of applications. *Logos* refers to any account, not just those for which a rational justification should, in principle, be available. It could be used of the tall tales that storytellers (*logographoi, logopoioi*) relate. One of the first Greek historians, Hecataeus, contrasts his account with the ridiculous tales the Greeks tell, using the term *logoi* for these (Fr. 1). But for his immediate successor, Herodotus (II 143, cf. V 36, 125), Hecataeus himself is (just) a *logopoios*, now principally in the sense of the maker of unreliable stories.

On the other side of the dichotomy, too, *muthos* sometimes means story, with none of the pejorative undertones of 'myth' as fiction. But when Thucydides (I 21) talks of the 'myth-like' quality of some earlier historians' accounts, which are beyond verification or scrutiny, we clearly already have some of those critical undertones.

Plato's usage is particularly revealing since he can use the two terms interchangeably, as he does concerning the 'likely account' (eikōs muthos, or eikōs logos) that he gives of the physical world in the Timaeus. But when he talks of the immortality of the soul in such dialogues as the Gorgias, Phaedrus and Republic, he recognizes that these 'myths' will not necessarily be believed by everyone. The contrast, there, is with logoi in the sense of accounts that are given a rational justification and will necessarily convince. Thus concerning the Forms, he says at Timaeus 29b5ff., one should give irrefutable accounts 'so far as it is possible', given that the subject-matter they deal with is stable and discoverable by the aid of reason, Being itself no less. The ideal is there stated quite clearly: yet one has to add that good-looking examples where Plato delivers on the implied promise are in short supply.

Aristotle, who is happy enough to use the term muthos for the plot of a tragedy, for instance,[4] nevertheless rejects any account that savours of the 'mythical' for the purposes of proper philosophy—just as he did the metaphorical. That includes Herodotus, who is labelled a 'mythologist' (muthologos) for what he says in his account of Egypt concerning the fertilization of fish. Aristotle himself is confident that there can be no question (as Herodotus II 93 had suggested) that female fish are fertilized by swallowing the milt of the male: their internal anatomy shows that that is impossible (On the Generation of Animals 756b6ff.).[5] But a text in the Metaphysics shows that, when it suits him, Aristotle is ready to adapt a traditional idea, even one entangled with mythical elaborations. At Metaphysics 1074a38ff., he reports a tradition which he says has been handed down in the form of a 'myth', to the effect that the heavenly bodies are gods and that the divine encompasses the whole of nature. But he distinguishes what he is there prepared to accept (namely that the heavenly bodies are divine) from the accretions about anthropomorphic and zoomorphic deities, which he says were added 'to persuade the many'

---

[4] Poetics 1451a16ff. Aristotle also uses muthos and logos interchangeably with regard to what we call Aesop's fables, Meteorology 356b11, Rhetoric 1393a30, b8ff.

[5] Aristotle comments that it never strikes those who hold such an idea that it is impossible, 'for the passage whose entrance is through the mouth goes to the stomach, not to the uterus, and whatever goes to the stomach is necessarily turned into food (for it is digested), but the uterus is evidently full of eggs: so how did they get there?'

and for reasons of political expediency; he means to have people fear divine retribution.[6]

Thus far the main thrust of the use to which the Greek philosophers put the category of 'myth' is to downgrade those accounts that did not meet the highest standards of clarity and rigour that they laid down, or that they simply decided did not seem reasonable to them. Once again, we may note that in Chinese antiquity, down to the end of the Han dynasty, there was no exact equivalent. That did not mean that Chinese writers accepted every story at face value. Tales concerning the remotest times and to do with the activities of legendary beings or cosmic giants are sometimes criticized as incredible. When the historian Sima Qian, for instance, reports some of the traditional stories concerning how dynasties were founded by women made pregnant by swallowing an egg or walking in the footsteps of a giant, he does so without any deflationary comment.[7] But when, in the continuation of one of those legends, he comes to the marvellous exploits of Hou Ji, the Lord of Millet, the patron deity of agriculture, he sees those as the result of hard work and skill, not of any superhuman powers he was supposed to possess. Yet Sima Qian does not deploy a category of the 'mythical' to do this undercutting: the Chinese term later used to translate 'mythical', namely *shen hua*, 'spirit talk', was introduced only in modern times as a borrowing from Japan.

But in the anthropological literature, 'myth' takes on an altogether different resonance. It has generally been associated with 'sacred tales' and there was a good deal of debate concerning the relation between these and legends, folk tales, and the like. But the use to which the category was put was transformed in the hands of Lévi-Strauss. His magnum opus, the multi-volume *Mythologiques*, is a virtuoso working out of the theme of myth as the vehicle for the expression of abstract ideas in concrete terms. The focus was on the underlying structures of narrative, not just in single myths, but in whole series of them, across languages and societies and continents indeed. The

---

[6] Aristotle was not the first to suggest that certain religious beliefs had been introduced for reasons of social control. The most notable earlier example comes in the *Sisyphus*, a play by Critias in the fifth century BCE, where a speaker says, 'I believe that a man of shrewd and subtle mind invented for men the fear of the gods, so that there might be something to frighten the wicked even if they acted, spoke or thought in secret' (Guthrie 1969: 243).

[7] See Lloyd 2002: 7–8 on the stories of Jian Di and Jiang Yuan, the mother of Hou Ji, in Sima Qian, *Shiji* 3: 91.1ff., and 4: 111.1ff.

reader was guided to the discovery of the underlying structure, often conveyed in terms of contrasts and oppositions, below the surface detail of the sequence of events narrated.

This suggested an altogether more fruitful way of exploring what can be learnt from myths, for he showed how they can be mined for cognitive content and should not be dismissed as childish fantasies. However, it has always been a problem to say how far Lévi-Strauss has strayed beyond what his subjects themselves would recognize or at least not disown, not least when they only have access to a small portion of the total corpus of myth under review. But in his more Platonic modes that would not deter him since his ambition is to reveal universal structures of the human mind. At that level of abstraction the difficulty was—as a telling critique in Sperber (1985): ch. 3 showed—to say what would count to confirm or disconfirm the hypotheses proposed. This moved the interpretation of myth beyond the unfruitful puzzlement over whether individual terms were used literally or metaphorically. But the price paid was to proceed at a high level of abstraction that tends to presuppose universal cognitive structures and thereby finesse the problems of the diversity of ontologies that have been central to my discussion here.[8]

Thus far I have examined certain relevant features of the use of language. But it is now time to turn to a further dimension of the problem, namely the impact of the audiences that speakers or writers address,[9] the contexts of communicative exchange in that sense as I have called them elsewhere (Lloyd 1990). It is obvious enough that it makes a difference whether a story is being told to a child, or as a joke, or in a religious ritual, or in a theatrical performance, or to a judge in a law-court, or to political authorities in support of a policy you are proposing or in the hope of gaining their favour. But my overall argument in this section is to use the examples of ancient China and Greece to show how the contexts of presentation may influence the

---

[8] Cf. also Wagner's criticisms (1978: 52) that the Lévi-Straussian analysis is limited to the cognitive contents of myths.

[9] One of the problems about the written word that Plato identifies is that there can be no interaction between author and readership. Books cannot answer questions. This is, of course, a point that Plato, the past master of dialogue form, makes in a written text, the *Phaedrus*, 276b–277a. A further problem with the written word that Aristotle among many others mentions is that it is not adequate for the purposes of instruction in such fields as medicine and navigation, where book-learning has to be supplemented by practical experience (e.g. *Nicomachean Ethics* 1181b2ff.).

contents of what is presented. There is a wealth of information on the subject in our sources, but we can at least hope to identify certain prototypical situations that will throw light on the issues.

Long before the unification many Chinese philosophers were so-called 'itinerant advisers' or 'wandering persuaders', *you shui*, who travelled from state to state offering counsel to rulers or ministers. This is what Confucius himself did, though he never found a ruler worthy of his advice. Mozi too was similarly involved. Mencius is recorded in audience with kings. Xunzi, Hanfeizi, Hui Shi, Gongsun Long all aimed to influence the policies of those in authority whether or not they held high office themselves.[10] Two of the great cosmological syntheses from the third and second centuries BCE were organized and put together by prominent statesmen. The *Lüshi chunqiu* was compiled by Lü Buwei, who was prime minister to the man who later became the first emperor, Qin Shi Huang Di, and the *Huainanzi* by Liu An, who was king of Huainan when his nephew Wu Di was emperor. Wang Chong in the first century CE had the same ambition to hold high office, even though he never succeeded in being appointed to a position of importance.

So some prominent writers were themselves important statesmen, and many more addressed rulers or ministers, offering advice on government. They construed their major task as being to secure good order in the state, and this meant being prepared to reprimand rulers and criticize them to their face, even if that was at considerable risk to themselves. Moreover, this model, of rulers interacting with advisers, spreads into other genres of writing. In the classic medical canon, the *Huangdi neijing*, much of the content of the treatise is conveyed in the form of reported dialogues between the Yellow Emperor, Huangdi himself, and his—medical—advisers. We are to understand first, that doctors have the ear of rulers, and secondly, that rulers would do well to listen to their advice. In mathematics, too, the first extant commentary on the *Jiu Zhang Suan Shu*, the *Nine Chapters on Mathematical Procedures*, by Liu Hui in the third century CE is careful to locate the investigation of mathematics in the broader

---

[10] Our indirect evidence provides us with information on the role of Mozi, Hui Shi, and Gongsun Long, even though in each case the writings ascribed to them are fragmentary, often of dubious authenticity, and subject to severe problems of interpretation. See Graham 1978, 1989, Johnston 2010.

context of the study of *yin* and *yang* and the *Dao*. That is a tactic he no doubt used to recommend his work to those in positions of power and influence. In his view the subject is of fundamental importance to everyone, even though he does not explicitly target an emperor.

Thus much Chinese theoretical writing is set in the context of a real or an imagined audience with a ruler. It is not just that this gives a certain tone of seriousness to the discussion. Just as good government is the recurrent preoccupation of philosopher-statesmen, so in most other areas of investigation there is a sense that the work should be of practical utility, that it should contribute to the welfare of 'all under heaven' indeed. At one point (ch. 5: 168.3–4) where Liu Hui discusses the properties of an abstruse geometrical figure, he feels the need to apologize: the figure in question is, admittedly, useless, but its study can contribute to the wider purposes of his mathematical inquiry.[11]

Of course classical Greek philosophers and scientists also went all out to persuade their target audience, but this was not a Sage King nor an emperor, but rather their peer group, if not the citizen body as a whole. In practice, in the political and legal domain, Greek statesmen regularly faced large numbers of their fellow citizens. The fickleness of their judgement is the topic of sustained adverse comment from the critics of democracy, who are far more fully represented in our extant sources than the advocates of democracy. Their ideas have generally to be reconstructed from accounts in hostile commentators or inferred from the actual policies they implemented and the political institutions they set up. But it was all very well to complain about the vagaries of popular judgement. Orators, if they were to succeed, had to learn to live with the situation, and they received advice on how to go about this in the numerous treatises on *Rhetoric* that were composed with that in view. They notably included Aristotle's three books on the subject, where due attention is paid not just to techniques of argument, but also on the need to know your audience and evaluate its character. It is obvious that such a skill, while not limited to cynical, manipulative purposes, could indeed be put to such a use, even though that was not Aristotle's aim.

---

[11] The figure in question, called a *bienao*, is a pyramid with right triangular base and one lateral edge perpendicular to the base, which Liu Hui uses in his investigations of the volumes of pyramids (cf. Lloyd 1996a: 152f., 161).

Outside politics, as well, those who competed as Masters of Truth for a reputation for their expertise in any number of subjects, from philology and mythology to medicine and mathematics, also generally had to face a public audience. One of the ways they built up a name for themselves—and made their living in the process—was by giving exhibition lectures, known as *epideixeis*, which would often be followed by question and answer sessions. Another was in public debates where rival experts confronted one another, and where it was sometimes the audience who decided, by their applause, who had the better of the argument.

Many writers draw attention to the parallelisms between political and sophistic debates. In Thucydides (III 37ff.) Cleon at one point reprimands the Athenian assembly for being easily misled. They all fancy themselves as connoisseurs of argument, but they do not treat the important business of politics as seriously as they should, but rather behave like those spectating at a performance of sophists. This is a nice touch not least because elsewhere Thucydides represents Cleon as himself an irresponsible manipulator of the assembly.

Plato's reaction to this was extreme. Socrates is made to say that he is not concerned with persuading all and sundry: the sole judge of his view is to be the Truth itself, which is never mistaken. Yet the very hostility of Plato's reaction may be taken as evidence of how pervasive the model of public persuasions in front of a general audience was. Again, as in China, the model extends into such areas as medicine and mathematics. Several of the Hippocratic works refer to public debates, including on subjects such as the constitution of the human body, where it was sometimes, we said, the lay audience who decided who had won, even though the subject—we should say—was quite a technical one. Some of our extant treatises indeed may actually derive from *epideixeis*.

In mathematics, too, Netz (2009) has recently assembled the evidence for what he dubs 'ludic proof'. He focusses especially on the Hellenistic period, including the mathematics of Archimedes, usually thought of as a paragon of seriousness. But, as he recognizes, already in the extant remains of mathematics of the classical period there are signs of a certain playfulness. That may be said to emerge in the pursuit of problems such as the squaring of the circle or the duplication of the cube, where mathematical interests are combined with a sheer delight in the virtuosity that can be shown in solving conundrums.

To double a cube was posed as a challenge by an oracle about the construction of an altar, at least according to one traditional account.[12] To rid the Delians of the plague that afflicted them, the god commanded that an altar double the size of an existing one be constructed. It is not just a mathematical problem, but a riddle to be solved on a par with the riddle of the Sphinx to which Oedipus alone had an answer. By Archimedes' day, the mathematics is more difficult, and the imaginativeness with which it was presented even more arresting. Who would have thought of tackling the problem of arithmetical notation by way of asking the question of how many grains of sand the universe would hold?

The Chinese certainly did not lack a sense of humour, but where advising the emperor was concerned, they certainly were not encouraged to waste his time. The contrast in classical Greece is not just that there was no emperor to persuade, but rather your fellow-citizens: we even have the sophist Gorgias offering as his advice that you destroy your opponent's seriousness with laughter, and their laughter with seriousness.[13] But then did he mean that seriously? A similar question arises with many of the wilder speculations of the Greeks, including the proliferation of ontological theories I reviewed in chapter 3, notably Parmenides' denial of change and plurality and Heraclitus' insistence on constant change. The context of presentation is the rivalry between competing Masters of Truth. But how seriously were their proposers committed to the conclusions they suggested? The arguments were serious challenges: but you could not *live* by their consequences, not in the case of Parmenides at least.

But we must now ask how the types of consideration we have been discussing so far in this chapter can be brought to bear on the problems of ontologies and the relations between language and reality. Defensively one may first express a note of caution. Evidently the language and contexts—including the audiences—of the presentation of the ideas in question need to be taken carefully into account, though this is sometimes difficult when the ethnographic reports are not explicit on those points. How and from whom were the ideas suggesting the underlying ontologies collected? There is a well-known problem

---

[12] Theon of Smyrna 2: 3–12, cf. Eutocius in his commentary on Archimedes *On the Sphere and Cylinder* 3: 88.23–90.4, both citing Eratosthenes as their source.
[13] As reported by Aristotle (*Rhetoric* 1419b2ff.).

of how far any particular informant conveys views to which the collectivity as a whole is committed (a problem for any society, our own included). As I noted before (ch. 2 pp. 43–4), the word of shamans is often doubted by the members of the society to which they belong, though this would not be because they were under the influence of hallucinogenic drugs—an accepted way of communicating with the spirit world. When they, or other prominent, or indeed ordinary, individuals report what they have seen, are they—or how far are they—to be trusted? In some cases the ethnographic reports make it clear that what is presented as what is to be believed and to be done may vary with the situation of the persons to whom the presentation is made.[14] As Barth (1975) showed for the Baktaman, each member of that society passes through a complex sequence of initiations, comprising six or seven stages, where what they are told at one stage may be flatly denied at another and where what they are told to do at one point is later revealed to have been a serious breach of taboo.

That underlines the obvious need for us to be careful, but does not help with the substantive problems. More positively we may go back to the dilemma posed by the literal/metaphorical dichotomy and attempt an alternative approach, namely by way of the notion of 'semantic stretch' that I introduced in Lloyd (1987: ch. 4). What that idea owes to other analyses and where it differs from them alike need some spelling out. It has often been said of metaphor that it calls forth or creates a similarity as much as it presupposes one (e.g. Black 1962). To call a king the shepherd of his people was ipso facto to make the king more shepherd-like and conversely shepherds more king-like. It allowed kings to be thought of as incorporating some of the properties of shepherds, protecting their people as a shepherd protects his sheep from predators, but also for shepherds to be imagined as king-like since, whether or not they owned their flock, they were in charge of and responsible for their sheep. Yet the interactive view stays, of course, within the framework of the literal/metaphorical dichotomy and as such still runs into the two chief difficulties it faces. First, there is the boundary problem, of saying where the literal ends and the metaphorical or figurative begins. Secondly, there is the dilemma: where the literal interpretation of a term strains belief, and the

---

[14] This tailoring of responses to different interlocutors is a feature both of Confucius' reported conversations as they are presented in the *Analects* and of Plato's dialogues.

alternative is that it is metaphorical, what is the metaphor a metaphor for?

To get round those problems, we need to generalize the interactive view, to allow that within any complex statement there will always be an interaction between any one term and the others in the collocation in which it is used. Already at the level of sentence meaning—and leaving aside any considerations to do with the speaker's utterance meaning (Grice 1968)—the syntagmatic lexical relations of each term modify, though of course they do not fully determine, interpretation of the collocation as a whole. What I call 'semantic stretch' has often been remarked upon under the rubric of 'polysemy' (e.g. Ziff 1972)[15] or in connection with the notion of 'semantic field' (Porzig 1934), but these have most often been considered features of particular lexemes, rather than a pervasive characteristic manifested to a greater or lesser extent throughout all natural languages. Similarly how far Wittgenstein's notion of 'family resemblance' applies beyond the paradigmatic examples of terms such as 'game' that he discussed, is controversial (cf. Hesse 1974: ch. 2, 1983: 28). Semantic stretch offers a radical alternative by sidestepping the boundary problems that all those dichotomies—between the literal and the metaphorical, between myth and rational account, between the playful and the serious—explicitly or implicitly presuppose.

But to suggest that every term has some stretch is not, of course, to say that it can mean anything, that its sense is indefinitely stretchable and its reference impossibly indeterminate, which would have the consequence that language is irredeemably vague or opaque. For a term to be used correctly, meaningfully, it must have a positive, interpretable relation to other uses both of that term and of those in the collocations into which it enters. In the expression 'to a green thought in a green shade' we are invited to engage in an exploration of the full range of our sense of what may be green. And our sense of what may be green grows once we are familiar with that verse of Marvell's, where his immediately preceding line, 'annihilating all that's made', just increases the challenging and provocative character

---

[15] Ziff 1972: 75 gave 'drive' as an example, pointing to the range of meanings, including 'backhand drive' in tennis, a 'drive' in the country, a cattle 'drive', an army's 'drive' on a particular front, sexual 'drive', and a garage 'drive'.

of his statement.[16] But if we are used to this in poetry, my proposal is that we should recognize semantic stretch far more widely, in communication in general and especially in heightened discourse including in such fields as ontology. We should, thereby—it is suggested—escape the stranglehold of an insistence on the literal as the sole proper use of terms. We should resist any idea that for something meaningfully to be called green it must be analysable in terms of a certain wavelength of light. In that example that point may be easy to grant: its possible application to ontology and science has generally been missed.

What lessons does this suggest for our strategic problems with ontologies? Let me go back to some of my earlier examples, to 'water', Greek *hudor* and Chinese *shui*, and to 'fire', *pur* and *huo*. A conventional view would have it that Greek *hudōr*, Chinese *shui*, and English water all refer to the same stuff, which chemistry analyses as $H_2O$.[17] To say that ice is water and that so also is steam is to say that all three are $H_2O$, though if we think liquidity is a defining characteristic of water, ice and steam are excluded.

Thus we have three natural language terms and one chemical formula. But we should not say that only the last is fit for purpose, and the others are all more or less potentially misleading. Rather, what there is for the terms to relate to is not a single item, but complex—multidimensional, as I put it. While there is an obvious overlap between the referents of *hudōr*, *shui*, and water, each of those terms has its own characteristic semantic range. Chinese *shui*, as we are told, is 'soaking downwards'. The focus is on processes. *Shui* conquers 'fire' and is conquered by 'earth'. 'Metal' produces it and it produces 'wood' and it is associated with a wide range of other items, the moon, the planet Mercury, winter, the North, shellfish, a

---

[16] The list of terms that may be associated with the concept of 'green' in *Roget's Thesaurus* is fourteen items long, namely 'new', 'young', 'pleasance', 'grassland', 'vegetal', 'sour', 'credulous', 'ignorant', 'unhabituated', 'immature', 'unskilled', 'pleasure-ground', and 'innocent', as well as 'greenness', where a further range of nouns and adjectives is displayed with a cross-reference to an even richer entry under 'vegetability', 'plant'. It so happens that Quine 1960: 127 sought to save 'green' from vagueness by reference to the comparative 'greener than', while still staying, he would no doubt have claimed, with the non-figurative uses of the term. But semantic stretch rejects a hard and fast distinction between 'figurative' and 'non-figurative', and allows interaction across the entire spectrum of uses of 'green'.

[17] The complex process by which it was concluded that this is indeed the correct formula for water has been well studied by Hasok Chang (Chang 2012).

particular musical note, and many other things in a series of analogistic correspondences that provide evidence of exceptional stretch. Even though Heraclitus chose water as a paradigm of flux ('one cannot step into the same river twice'), for most Greeks *hudōr* is not a process, but a substance, an element indeed. But in their theorizing many different kinds of substances are 'water', not just hail, and ice and frost, but also metals. Plato explicitly distinguishes between two main kinds of 'water', the liquid and the fusible (*Timaeus* 58d), the latter including the metals. Aristotle, who is generally so critical of improper linguistic use, is happy enough to consider not just metals, but glass and certain stones as 'belonging to water' (*Meteorologica* 389a7ff.). But rather than *just* dismiss Aristotle's view as a typical piece of mistaken archaic science—now superseded by modern knowledge—we should at least allow that his analysis of *hudōr* shows that he puts a greater emphasis on liquefiability or fusibility than is the case with our common understanding of 'water' in English.

The first temptation to be resisted here is to think that we can set limits to the 'literal' use from which others deviate, even when, as with Aristotle, the literal/metaphorical dichotomy is available as an analytic tool. But the converse mistake is to think we are dealing with mutually unintelligible languages. That will not do, because the semantic range in each case can be circumscribed at least in broad-brush terms—and that is the mode of circumscription that is appropriate here. The senses and referents of the relevant terms in English, Greek, and Chinese differ, but overlap, not that there is a neutral language in which to parse this. Translation is indeed provisional and indeterminate, though not radically so in the way Quine suggested. We should not imagine that discourses are hermetically sealed off from one another, any more than that the ontologies in question are totally mutually incomprehensible.

A similar analysis can be given where 'fire', *pur*, and *huo* are concerned. Fire was seen as a substance by Empedocles, Plato, and Aristotle (though Heraclitus saw it as ever-changing and constitutive of the cosmos as a whole), while *huo* is a process in *Huainanzi*. Both views invite one to focus on and appreciate certain characteristics of what burns. Despite the fact that for centuries many Europeans accepted that fire was an element, we now find the substance ontology more difficult to come to terms with in this case than the notion that fire is some kind of process. Yet pondering ontological alternatives is not a matter of accepting just one, that is, choosing between them, so much as understanding them, and that means, for example, following the chain of ideas that led to the supposition that fire remained a part of the compounds that it contributed to creating. What we mean by fire is not to be limited to what a dictionary may give as a definition: nor (to repeat myself) is it helpful to attempt to fix the limits of the literal and the figurative. But nor is it the case that we can use the term quite arbitrarily of anything we like. Who would want to do that in any case? Meaning has to be built up by considering whole collocations of terms, the interpretation of each one of which may be affected by the changing interpretation of others as the exploration proceeds.[18] This applies also to the development of scientific vocabulary.[19] Even when neologisms are created, their use has to be explained in language that draws on, even if it may substantially modify, existing meanings. The pragmatic ground-rules of communication—relevance and sincerity in particular—remain,[20] even though the playful may intrude on the serious.

So where does that leave us on the question of how language connects to the world? Words and things belong, to be sure, to different registers, and the common idea that the one can be used to map the other is misleading, since there is no independently accessible 'other' for words to map on to. So a demand for such a connection, let alone one that should yield a one-to-one correspondence, is misplaced. Reality does not come ready packaged for us merely to attach our concepts to those packages: rather we create the divisions, not that that means (as I have said before) that they are quite arbitrary. Both relata are complex, for on the one side the terms we use are subject to stretch, on the other what there is for them to pick out is often

---

[18] Thus semantic stretch shares with the idea of charity of interpretation that I discussed in chapter 2 that it warns against imposing our—observers'—categories on the discourse of the actors we are interpreting. But where charity assumes that that discourse has eventually to be assessed in our terms, semantic stretch suggests that our categories may need revising to take theirs into account.

[19] This is so even in the cases Kuhn insisted on, where a whole group of key concepts undergoes radical revision. I shall be returning to this in chapter 5.

[20] Grice 1975, 1978 introduced nine 'maxims of communication', covering sincerity and relevance among other principles, and classified these in four categories, of quality, quantity, relation, and manner. Among the extensive literature that this occasioned, see especially Levinson 1983: 101ff., Sperber and Wilson 1995: 33–6.

multidimensional. But the lack of a simple correspondence between words and objects obviously does not mean that communication 'about' things—whether verbal or non-verbal—and indeed acting 'on' them are alike impossible. Linguistic communication is in any case generally a matter not just of the semantics of single words, but of their use in collocations. That is certainly subject to error and failure, as I discussed in chapter 2, but not inevitably so. Our interlocutors often confirm that they have understood at least up to a point, though mismatch between what they understood and what we intended is always possible. Again a mistaken appreciation of the practical situation we face may mean that our interventions fail to achieve their desired result. While both these activities, communication and action, may misfire, the very fact of possible *error* suggests that we can be more, or less, successful in both, and that would not be the case if all attempts were equally beyond our capabilities.

Adopting a semantic stretch analysis means renouncing the literal/ metaphorical dichotomy and at the same time escaping from the straitjacket that it tends to impose—as does the demand that we communicate by means of well-formed formulae. Univocity thereby becomes an exceptional or limiting case. To be sure, geometers need clear definitions (though not necessarily per genus et differentiam) of 'circle', 'line', 'straight', and so on, for their investigations, but that fact should not mislead us into neglecting the semantic stretch—the multiple meanings and associations—of those very same terms outside the specialist context of geometry, though, to repeat my earlier point, there are certainly limits to that stretch in all three cases.

That analysis (it seems to me) is better suited to ontological exploration—we may allow water to be a substance *or* a process—and to the recognition of the multidimensionality of reality: colour is not either (just) hue or (just) luminosity or (just) saturation, but all three. Any language will offer multiple opportunities for capturing the differing aspects of whatever subject we wish to talk about. Whatever natural language they use, our interlocutors invite us to join in exploring how what they say can make sense, to us as well as to them (even though initially it may seem to us to be nonsensical or just plainly mistaken), and with live interlocutors at least some of our misunderstandings, including our mistranslations, can be corrected, even though, as I said before, neither translation nor understanding can ever be perfect. Of course some statements may be plain enough and do not generate the puzzles that I have been discussing. But any use of a general term already raises the question of the similarities between the items to which it refers and the differences from those to which it does not. In any but the most straightforward communication, interpretation will demand skill and will remain open, and that applies across the field from poetry to statements of values and to cosmologies.

It might be objected that such an analysis leads, nevertheless, to linguistic anarchy. How, the critic will protest, can a thought be green? Well, the answer might be, quite easily, when we reflect on the associations of green with youth and freshness and innocence and even naïveté. Given that the aspects of the phenomena focussed on can be separately identified, and that the range, or stretch, of the terms used to communicate them can be rehearsed—always using language to be sure—we can resist the objection, even though there is no neutral vocabulary in which those tasks can be carried out.

It is not my claim that this solves the puzzlements of alternative ontologies and the problem of the relations between words and objects at a stroke, that all the difficulties can be resolved by a judicious use of the notions of semantic stretch, of complementary styles of inquiry, and of the multidimensionality of the phenomena and indeed of reality. On the contrary, I accept that many puzzles remain, both strategic and particular. But I would treat diverging ontologies not (generally) as presenting alternatives between which a choice has to be made, with the object of identifying one as correct and the rest mistaken, but rather as challenges to our understanding, where with imagination we can learn from unfamiliar views and attitudes to question our own assumptions. The issue between a substance-based ontology (such as our own ordinary language tends to assume) and a process-based one (such as underlies much classical Chinese thought) is far from an open and shut case, and indeed continues to be disputed when modern science confronts our normal mundane perceptions. In the final analysis there is a lesson to be learned from that chapter of *Zhuangzi* where 'lodge sayings', *yu yan*, are recommended, that is, that we should put ourselves in the position of those we are talking to, to see things as far as possible as they see them.[21] That does not tell us what they see. But if both sides in an exchange take 'lodge sayings' as their model, there is hope for some instructive communication between them.

---

[21] Although Zhuangzi was concerned with communication with his fellows, not with understanding foreign peoples, his notion of taking up another's position meshes very well not just with the general anthropological recommendation to stay with actors' categories so far as possible before applying observers' ones, but also with the sense of the importance of allowing plural points of view that perspectivism draws attention to. Zhuangzi does not talk about jaguars, though I cited before, in chapter 2, his puzzling over the feelings of minnows, and in a famous passage (ch. 2: 94–6) he wonders whether, on waking from a dream about a butterfly, he is Zhuangzi who has been dreaming of the butterfly, or the butterfly now dreaming it is Zhuangzi. The text does not negate the difference between Zhuangzi and a butterfly, but it uses the story to illustrate what it calls the 'transformation of things'.

# 5

# Philosophical implications

MY aim in this chapter is to return to the major interpretative issues that have been my chief concern throughout this book, to examine how far some clarification can be achieved, and a resolution, if not a solution, proposed. Chief among the problems has been the conditions of possibility for mutual understanding. Ontologies evidently differ fundamentally, and not just in the ways identified by Descola's tetradic taxonomy and Viveiros de Castro's perspectivism. In what sense do they imply radical incommensurabilities? When conceptual schemata, cosmologies, and paradigms appear to have no common ground, when perceptions vary with who does the perceiving, how is mutual comprehension possible? If the worlds themselves are incommensurable, are those who perceive them, inhabit them, live by them, cut off from one another completely? Are we faced, in other words, with worlds between which no communication is possible? It is at this point that I have reservations, as will already have become clear. Comprehension is always difficult in the kinds of cases I have been dealing with, but if and when we can make some progress towards securing it, it is all the more valuable for that very reason.

Closely connected to that first issue of incommensurability, secondly, is the difficulty of securing objectivity in judgement. Faced with any problem of interpretation, in history, ethnography, philosophy, or science, how can we do anything other than apply our own conceptual categories, preconceived ideas and values? Do we not get out simply what we have ourselves put in? Are we not locked in a barren hermeneutic circle of our own devising?

Then there is the third fundamental problem of truth itself, the subject of so much philosophizing, including notably in the twentieth century. Should we adopt a correspondence theory of truth, or one that settles for consistency? If the former, how are we to imagine that access is to be gained to what it is that true statements are held to correspond to? Unmediated access seems impossible. On the other hand, the criterion of consistency seems quite inadequate, for it is clearly the case that sets of statements can be internally consistent but all false.

Fourthly, there is the seemingly interminable dispute between various brands of realism and different types of anti-realism, relativism, and social constructivism. Does science yield, in principle, a correct account of external reality? Or is it just what a particular community decides is worth studying (where that may evidently change over time and with the group concerned)? True, as already noted, like 'realism', 'relativism' is not a single, well-defined philosophical, scientific, or even commonsensical position, but can be invoked in a wide variety of contexts, many of them quite straightforward and uncontroversial. Thus any use of comparative adjectives, such as hotter, colder, implies that the instance of the quality in question is being judged relative to some other instance or norm. But in the context of science, relativism usually stands first, for the rejection of any absolute claims to truth or access to ultimate reality, and secondly, for the insistence on relating any theoretical claims to the point of view of those who make them. While the first is comparatively unproblematic, the difficulty with the second is that it may be invoked to allow just any set of beliefs to be entertained and to be treated as equally well justified.[1]

Let me leave to one side, for the moment, the first of those problems, but deal with the other three to examine what light our preceding studies can throw on them.

First, then, to the problem of objectivity, which has been a recurrent preoccupation of the sociology of knowledge (e.g. Collins 1998): its recent history in European thought has been the subject of a detailed study by Daston and Galison (2007). Objectivity generally implies a notion of facts of the matter that are independent of any subject or person apprehending them. But how can we arrive at any account, whether in historical or ethnographic studies, that is not merely a reflection of our own prior beliefs? We cannot, I said, on pain of distortion, impose our own conceptual framework on others' ideas.

---

[1] See Smith 2011 and the other contributions to the 2011 issue of *Common Knowledge*.

Yet we have to, since our conceptual framework is the only one we have. I have to translate Greek or Chinese into English. True, we sometimes do not translate, but transliterate. Ancient Greek *logos*, which spans word, account, reason, and much else besides, is untranslatable, and so too are Chinese *qi* (breath, energy) and *yin* and *yang* (the negative and positive principles). But we cannot interpret ancient Greek simply by using Greek. Sooner or later we have to leave the ancient language and use a modern one, in my case English, with all the potential distortions that may bring. So the problem remains.

But while it is true that we have no option but to use our own conceptual framework, we can evade the apparent dilemma if we bear in mind that that framework is no monolith. We acquired our conceptual categories in the complex processes of education we underwent as children—and that included being taught science and becoming acquainted with other languages besides one's own mother tongue. We were introduced—we continue to be introduced—to new and unfamiliar ideas which present more, or less, severe challenges to our understanding. We may, to start with, not get the point. Some people are quicker at picking up languages, or learning mathematics, some slower. But my point is the simple one that learning whatever mathematics we learnt at school gives us plenty of examples from our own experience where we recognize without difficulty that we acquired not just new knowledge, but also new conceptual tools.

The example of language learning may be helpful in another respect. However good I may become at ancient Greek or classical Chinese, I am never going to be the equal of those who spoke those languages in ancient times. I can never identify with them, to 'put myself in their shoes', as we say, in the sense of sharing all their assumptions, preoccupations, values. We can never master any ancient language totally—and the same point applies, of course, to modern ethnographers acquiring the language of the peoples they study. But then we should also ask ourselves the question of whether we can ever master even our own mother tongue *totally*. What it means to say that we have a 'complete' command of English is that we are fluent in British, or it might be American, English. But very few people could claim to be equally fluent in Caribbean and Indian and African English as well, let alone the English dialects spoken in Glasgow, or in parts of London, or in the Fens outside Cambridge.

What this means for the general problems of objectivity and inter-pretation is this. We evidently have no option but to approach the problem of understanding others with the conceptual framework that we have. But that does not mean that we have to stay with it, nor that it cannot be revised. Quite how we learn new ideas and skills, and revise basic assumptions, may be difficult, even impossible, to charac-terize in abstract terms—and obviously not everything can change all at once. Developmental psychologists, cognitive scientists, philoso-phers of mind, have plenty to disagree about, concerning the stages of children's normal development, about the processes by which they come to revise the naïve physics or the naïve psychology that they originally assume, about the modules that are needed to account for those, or even about whether the assumption of modularity is correct. That has been the topic of ongoing controversy ever since Fodor first raised the issue.[2] But a timely reminder *that* learning occurs should encourage us in the belief that it can continue to occur, that we can continue to expand our horizons, modify our preconceived ideas and values, and, yes, even some of our basic conceptual categories, as when we may come to appreciate, for example, that a process-based ontol-ogy is just as viable as a substance-based one. Indeed this is precisely where the study of the systems of beliefs of other societies, ancient and modern, is so valuable. It introduces us to possibilities that were initially beyond our imagination. Some, to be sure, have more evident potential than others: but then it is up to us, using, precisely, our imagination, to see how fruitful each may be.

That is not to say that there is any guarantee that our efforts will be rewarded with success. As we all know, it is far easier to stay with our own opinions, values, habits, to fail to do justice to what is truly new when we encounter it, especially when we may suspect that it poses something of a threat to our own familiar, cosy, assumptions. But if there is no guarantee of even modest success, there is no reason to believe in its impossibility either. Warnings as to the difficulty of achieving a proper degree of self-awareness and of practising

---

[2] The literature on modularity, following up Fodor 1983, has been vast. With, for example, Barkow, Cosmides, and Tooby 1992, Sperber 1994, 1996, Carruthers and Chamberlain 2000, Atran 2001, compare the doubts and criticisms expressed, for instance, by Karmiloff-Smith 1992, Tomasello 1999, Panksepp and Panksepp 2000 and Sterelny 2003. The literature on childhood learning is equally extensive. For a recent overview of many aspects of the question, see Carey 2009.

self-criticism are always valuable. But we do not need profound philosophy, we just need to remind ourselves of our childhood, to see that some progress in learning, in grasping new ideas and understanding other people, is not ruled out. If we can never achieve total, theory-free, objectivity, there is no need to believe that everything is a matter of arbitrary subjective judgement. That will not satisfy the hard-line post-modernist, to be sure, but the problem with the real hard-liners is to say what would satisfy them (cf. Runciman 2009: 220–2).

This takes us to the next problem, truth, and to the dilemma posed by the unsatisfactoriness of each of the two main competing theories of truth. The correspondence theory seems fatally flawed, since there is no direct access to a reality 'out there', that is, no access unmediated by words—and as I have insisted, they are all more or less theory-laden. But if correspondence to reality is, then, strictly impossible, a mere consistency or coherence theory of truth is evidently not enough. It is clearly no good allowing just any set of statements or beliefs that are internally coherent to be, by that token, true, since there are plenty of such sets that are palpable nonsense. Nor will it do to allow what is accepted by some group, maybe even some group of experts such as scientists or philosophers or elders or ethnographers, to count as true, since the invocation of experts does not make the objection go away. Appeals to authority still leave open the question of the grounds on which the authority claimed is based.

The trouble with overarching, and by that token monolithic, theories of truth, as I already argued in Lloyd (2004: ch. 5), is that they seek a single account that will be valid across the board. But reflection on how we use the term 'true', and what we need it for, suggests that that may be the wrong approach. An analysis of the ancient historical data is doubly revealing, first, because some commentators have gone so far as to deny that the classical Chinese had a concept of truth, and secondly, because we can trace the origins of our modern philosophical problems to ancient Greece.

The Chinese, to be sure, did not propose the types of overall theories of truth that we find among the Greeks and that persist with us today. But there are at least three contexts in which the Chinese, already in the period before major Buddhist influence, made considerable play with related ideas.

First, the Chinese had no difficulty in assigning truth values to statements. The commonest way of doing so is by recording that things are so (*ran*), or not so (*bu ran*). Secondly, the antonyms *shi*, and *fei*, are used to mark the contrast not just between right and wrong, but also between what is and what is not (cf. Harbsmeier 1998: 194, Lloyd 2004: 59). Thirdly, verifying that a claim is correct is a concern in a variety of contexts (in relation to legends, for instance), as also is assessing whether appearances are quite what they seem, in the matter of human character, for example. What is it to be a true king, *wang*, or a true 'gentleman', *junzi*? A great deal of Chinese philosophizing, from Confucius onwards, centres on such problems.

So both the contrasts and the similarities to the ancient Greeks are striking. First, as to the contrasts, the Greeks formulated not just one but three main overall sets of positions on truth. It is from these debates that our own controversies stem. They may be labelled the objectivist, the relativist, and the sceptical. Thus among the objectivists, Parmenides links truth with necessity. The Way of Truth starts with the statement: it is and it cannot not be. The subject ('it') is left unstated, but one possibility is that it covers whatever it is that you are about to investigate. We cannot investigate something of which we say it is not. How Parmenides believes he can justify the further claim of necessity—it is and it *cannot not be*—is even more controversial, though it is clear that that is the claim he makes.

Plato then unties the knot between truth and necessity. Truth is now a necessary, but not a sufficient, condition of knowledge, for he recognizes that opinions can be true: they are not all false. In the *Sophist* (263b), truth and falsehood are analysed as properties of statements: the first says of a subject things that are, the second other things. Aristotle, with a very different ontology, states that every affirmation and denial must be either true or false (the Principle of Excluded Middle) (*Categories* 2a7ff.). Yet both he and Plato continue to use 'true' as synonymous with genuine—so both continue to use true and false, *alēthēs* and *pseudēs*, of things as well as of statements.

But ranged against the various realist positions, there were those who disagreed with any claim that beyond the appearances there is objectivity to be had. Our chief evidence concerning the relativist position comes from Plato, unfortunately an implacably hostile witness. For Protagoras, he tells us, man is the measure of all things, of what is that it is, and of what is not that it is not. What appears hot to me is hot to me, even though it may appear cold to you (and so is cold to you), and as Plato develops that position, in the *Theaetetus*, the same relativizing principle applies not just to sensibilia such as hot and cold, but also to moral qualities, good, bad, just, unjust. Quite where Protagoras' own ideas end and where Plato's interpretation of them takes over is disputed, but that need not concern us here, because for us the interest is in the positions that were canvassed, not with who canvassed them.

The other main challenge to the objectivist position came from a variety of sceptical attacks. Some denied the possibility of knowledge concerning hidden reality or unseen causes altogether, though it was soon spotted that since such a denial had to be based on some criterion, it breached the very principle it sought to advocate and so was vulnerable to the charge of self-refutation. So other, Pyrrhonian, sceptics adopted the safer course of withholding judgement, *epochē*, neither asserting nor denying on matters concerning underlying realities or the hidden causes of things.[3] It may or may not be the case that the ultimate constituents of physical objects are atoms, or earth, water, air, and fire, or whatever. But there is no criterion on which to base a judgement. Perception is unreliable, but so too is reason. To opt for one or for the other itself presupposes a decision that cannot be taken neutrally, and so judgement should be suspended. Given the plethora of ontologies that we saw the Greeks propose (chapter 3) one cannot but feel some sympathy with those who thought the issues undecidable. Indeed they are not decidable, if by that one means the problem can be resolved by plumping for one answer to the exclusion of the others.

But what was needed, we may say, was not a bland or despairing suspension of belief, but first, the revision of the terms in which those ontologies were seen as in competition with one another and a choice had to be made between them, and secondly, a more sustained exploration of the considerations that were thought to tell in favour of each of them. It is those revisions and explorations that hold out the hope of some increase in understanding, not that all the disagreements and disputes go away. The aim is not to have every position come out true,

---

[3] Compare the rejection of both assertion and denial in *Zhuangzi* (above ch. 3 p. 66 n. 15), although that was not based on an epistemological argument by which both reason and perception were categorized as unreliable.

but rather to comprehend their varying strengths and weaknesses and to appreciate why they were entertained in the first place.

Those radically opposed Greek points of view on being and truth are well known. They provided the epistemological and ontological underpinnings of the competing cosmological, physical, and ethical theories that rival Greek Masters of Truth proposed. Yet not all Greek invocations of truth and true relate to questions to do with fundamental reality. Some have to do with more mundane matters—which is where some Greek data share certain points in common not just with the Chinese, but many other societies across the ancient and the modern worlds.

The vocabulary of truth, as used in ancient Greece by writers such as the historians and the orators, is very different from the abstract speculations of the philosophers. The important issues those writers discuss relate not to some ultimate criterion, but to such matters as the honesty of witnesses, the accuracy of reports, the diagnosis of intentions, and the like. Thucydides (I 23), for instance, announces that the truest cause, *alēthestatē prophasis*—as opposed to the alleged reasons—for the outbreak of the Peloponnesian war, was the fear that Athenian ambition inspired in the Spartans. Over and over again, in Greek forensic oratory, the defence and prosecution dispute the reliability of testimony, each side protesting that its account is the whole unvarnished truth, while the opposition's case is based on distortions, obfuscation, and downright lies. In the natural philosophers and scientists themselves, the *actual* procedures used to check or prove results are often more complex than their explicit theories on the subject of truth might lead us to expect, involving, not surprisingly, complex combinations of abstract arguments and empirical considerations.

The comparative material I have passed rapidly under review serves to suggest two important points. On the one hand, there is a question about origins. Where Greece is concerned, explicit formulations of rival positions on the theory of truth as such appear to be connected with a certain strident adversariality I have remarked on before in situations where Masters of Truth battled for prestige: most saw the need to back up their ideas on first-level issues with a well-thought-out repertoire of responses to second-order challenges, concerning the very basis on which they claimed superior knowledge in the first place. On the other hand, comparisons can, I believe, be used not just to trace the source of the issues, but to contribute to their resolution.

The questions to do with truth that we should not lose sight of include, as a central core, truth-telling (sincerity), authenticity, checking, and warranting (cf. Williams 2002). Common as these concerns undoubtedly are, no one is in any position to claim that they are universal across all human societies, and as to the values attached, we should definitely be wary of assuming cross-cultural uniformity there. We have only to think of the way in which, in certain contexts at least, the ancient Greeks positively admired those (like Odysseus) who were good at lying and deceiving.[4] They praised, under the rubric of *mētis*, cunning intelligence, the skills that included the ability to win by fair means or by foul—provided that you were not found out (cf. Detienne and Vernant 1978).

The contexts in which the veracity of speakers, or the reliability of witnesses, may be questioned, and the procedures that can be used to check them or other evidence, are all enormously varied and reflect very diverse practices and challenges. So too are the degrees of accuracy and of precision that are to be expected. To cite one of my favourite examples, the approximation that a builder can tolerate for the circle-circumference ratio differs from the goal a mathematician may set in determining its value. The notion of the approximately true is anathema to logicians, since it destroys the transitivity of entailment. But plenty of ordinary human calculations make do happily enough with just such approximations. In today's scientific investigations, verification takes forms undreamt of a hundred years ago, let alone in antiquity, wherever in the world you were, and with the increase in the range of what can (provisionally) be confirmed goes also a greater sense of what is beyond that range—and not just in the case of Heisenberg's uncertainty principle. It has often been observed that the advance of science is accompanied by an ever-increasing appreciation of just how much remains unknown.

The lessons I would draw from my comparative analysis are, then, as usual, pluralist ones. The variety of manifestations of the problems of truth calls, as I said, for context—or at least domain-specific responses, rather than the invocation of a single universal principle.

---

[4] At *Poetics* 1460a18f., Aristotle tells us that Homer taught the Greeks how to lie well.

Truth-telling has important and distinctive roles to fulfil, in law and history, as well as in science and philosophy, not to mention also in everyday life. The capacity of humans (philosophers especially) to doubt—which is where the challenge to validate a claim generally arises—is almost limitless, but that is not to say that it is always reasonable. That certainly does not mean that the search for truth is hopeless and has to be renounced—let alone that we have to look in an altogether new direction, such as art or religion, to find *the* key. That is more likely to add to our difficulties than to put us out of our misery, if, as I argue, there is no *single* key.

Realism and relativism, the next problem on my agenda, also take the form of an apparent impasse between two incompatible positions, between which, nevertheless, a choice looks as if it has to be made. Both capture some elements of the truth. But both run into fundamental difficulties when presented as global solutions to the problems. A certain naïve realism falls with the fall of the correspondence theory of truth. Relativism looks as if it can give an account of scientific change—but at a price. What was taken to be the best science a hundred years ago was very different from today's, and there is no reason to think that today's will not be superseded in its turn. Another obvious point that relativism captures is that scientific researchers are no more immune than anyone else to influence from the community to which they belong. So even scientific programmes are subject to such pressures. Yet that is not to say that such influences determine the programme as a whole.

That is one point at which relativism (as sometimes construed at least) may come unstuck. Another is that theories are revisable not just according to the already existing norms or fashions of the group. True, to claim they are modified to give greater conformity to what is the case—the truth—may sometimes be mere rhetoric or shorthand for a long and complicated story about the justification for the procedures used and the results obtained. But if the relativists protest that they must be given clear rules for how greater objective accuracy is to be obtained—or else that idea is to be rejected as incoherent—then my answer, as in the case of theories of truth, is to resist the question and to insist that procedures, tests, methods cannot be generalized over, but are always context-specific.

Comparative history and ethnography have lessons for us in this area too. We need to be clear, among other things, about what there is for theories to be theories of. We are concerned with understanding. But what is there to be understood? The answer to that, as I have urged, can certainly not be taken for granted. To underline the point again, let me revert to the problem of the classification of animals that I used as a prime example in Lloyd (2007). Implicit classifications of animals—and very different ones at that—are built into every natural language. The languages themselves give the appearance of laying down ideas on the genera and species of animals once and for all. Yet that may be deceptive. In most cases in the ethnographic literature we cannot document changes, since the diachronic evidence is limited or non-existent, but in both ancient China and ancient Greece we can. Certainly in China different classifications are proposed in such classical texts as *Huainanzi* and *Erya*. Both those texts give a fivefold classification of the main animals, but it is not the same five in each case: nor are the five classes associated with other fivefold groupings in precisely the same way.[5]

Where Greece is concerned, Aristotle certainly did not merely accept the notions encapsulated in the language he spoke. He challenged some of the implicit assumptions of that vocabulary and he was led to coin totally new terms for the new groups he identified, among the 'bloodless' animals especially. Yet neither *Huainanzi* nor Aristotle was trying to anticipate modern biology. Their programmes reflected their own, distinctive, cosmological, concerns. To be brutally brief, *Huainanzi* is concerned with what goes with what, the correspondences between things. Aristotle studies animals, as he tells us, mainly to investigate their forms and final causes. He has, then, teleological motivations, ultimately to show the beauty of nature.

Yet it is instructive to consider where modern science has now got to on this subject. First, there is what Mayr (1957) calls the species problem, where the permanence of species had, of course, already been overthrown by Darwin. Attempts to define 'species' independently of all assumptions fail. Which groups of populations should be given the rank of species? We can collect similarities and differences, but they have always to be *weighted* and that evidently risks circularity. In fact, many different criteria have been and continue to be invoked. Morphology and interfertility used to be two favourites, to

---

[5] I may refer to Lloyd 2007: ch. 3 for the details.

which we should now add groupings based on DNA analysis. But although some of the results using different criteria converge, that is far from true across the board. In particular, those three do not.

The lesson is that there is no one 'correct' taxonomy of animals—any more than there is of plants. There, above the family level, the orders of plants remain deeply controversial, despite the very considerable efforts that are made, including by international committees set up for the purpose, to impose standardization. But that does not mean that anything goes, as some versions of relativism or social constructivism might suggest. The fact that different schemas relate to different programmes does not mean that they cannot be judged objectively *with regard to* whatever programme they pursue. You can have more, and less, accurate classifications of animals focussing on DNA, for instance, or on morphology. True, when there is interference from symbolic systems, that complicates the picture, and the evaluation has to proceed by the criteria of what is appropriate or felicitous, rather than what is accurate. But recall that I have insisted that science itself is never totally value-free. The conclusion we should draw, in the classification case and in many others, is that reality is multidimensional.[6]

The second principle that allows space for alternative accounts of the same subject-matter is that of different styles of investigation, which I introduced in chapter 2, where I noted what my own application of the idea owes to earlier uses by Crombie and Hacking. For my purposes, studying the ancient world, 'styles' are constituted by groups of salient ideas, aims, methods, and modes of presentation. That can be illustrated by the contrast I have mentioned before, between the frequent Greek demand for a mode of understanding based on deductive demonstration, and the greater Chinese preoccupation with associative understanding, with the correlations between things. Yet that is not to suggest that a global contrast, between Greek thought as a whole and Chinese thought as a whole, will do, since, as I have explained, there are notable exceptions on both sides. It is not that all programmes are equally successful in their own terms. As I said, it is not that anything goes, not that I am claiming (of course)

---

[6] Multidimensionality applies at different levels depending on the way the subject of investigation is defined. Sometimes this is at the level of perceptible phenomena, sometimes at that of what is postulated as the underlying reality. Sometimes indeed such a contrast between phenomena and reality is hard to draw, or even irrelevant.

that we have some Olympian vantage point from which we can stand in judgement on everyone else.

So to treat realism and relativism as mutually exclusive and exhaustive alternatives is liable to mislead. In a sense, we need both. But in a stronger sense, we need neither as they are often presented. Rather, we need to assess, in every case, the interaction of received ideas and attempts to go beyond them, the extent and limits of the influence of peer groups, and the modes of the revisability of theories. We need to allow first for the multifaceted character of what there is for theories to be theories of—what I have been calling the multidimensionality of reality—and then for the diversity of styles of inquiry directed at different, or even the very same, aspects of the phenomena.[7] Once again we need to guard against the assumption that there are always simple and definitive solutions there to be found.

I return then to my final strategic problem, to do with incommensurabilities. This notion too comes in different strengths, as I said, and we need to distinguish the points that are acceptable and indeed obvious from those where we may raise objections. Sense and reference, the two concepts standardly appealed to in analyses of meaning, have both been heavily problematized in modern philosophical debate, notably in the wake of Quine's notions of the indeterminacy of translation (on which I have had some points to make in previous chapters) and of the inscrutability of reference. We may note, to start with, that the distinction between the two is far from transparent, however much some favourite simple-seeming examples might contrive to make that appear to be the case. The Morning Star and the Evening Star, it is said, have the same referent (the planet Venus), but not the same sense. But Venus is not, for astronomers at least, a star, so 'star' has to include other heavenly bodies.[8] We can allow that, according to the principle of semantic stretch, but if we probe the

---

[7] Thus these two ideas, multidimensionality and styles, combine elements of more conventional 'relativism' and 'realism' and thereby resist treating those as exclusive alternatives. Any inquiry will need to be judged relative both to the particular dimension of the subject investigated and the style employed. So thus far 'relativist'. At the same time there are objective criteria for assessing how accurately an inquiry captures what it is an inquiry into, and indeed the multidimensionality of the subject-matter is given independently of any particular style. So in those respects the assumptions are 'realist' ones.

[8] Of course, other standard examples to make the distinction are not so problematic. Cicero and Tully are indeed two proper names for the same individual.

question of the sense of 'Morning Star' and 'Evening Star', further problems arise. 'Morning Star' can be glossed as a particularly bright heavenly body (though ordinary stars will not be allowed to count) shining before the sun rises in the morning. But Jupiter is sometimes as bright as Venus and in the right position close to the sun in the dawn sky over a period of months. It can easily be taken for a 'Morning Star', and an exactly analogous problem arises with whatever is to be labelled as an 'Evening Star'. The indeterminacy I have pointed to in the senses of both compound names is set aside by the convention that the referent is one astronomical body, indeed (as the Babylonians knew long before the Greeks) the same one in both cases, the planet Venus. So it is an oversimplification to say that we have two names for the same planet, Venus, though Venus can, of course, appear now as one, now as the other.

That is not to say that the contrast between sense and reference cannot be made. But those two may interact. That applies particularly to the exemplary cases where the meanings of groups of key scientific terms are said to shift, as in the examples of force, mass, and weight exploited by Kuhn when he developed the notion of incommensurability.[9] What shifted, between Aristotle and Galileo, was both the referents of those terms and the senses by which they were to be defined. If when the two paradigms are labelled 'incommensurable' that refers to those shifts, we may readily agree: there were indeed major displacements in the concepts that were put to work. It is when incommensurability is made to carry the further, stronger implication that there is no possibility of mutual comprehension across paradigms that problems arise, for that flies in the face of the two points, first that Galileo had a fair idea of what Aristotle meant, and secondly that we should not think that either of their views is quite beyond the reach of our understanding, even though we have no totally theory-free vocabulary in which to express ourselves.[10] It is certainly difficult to

---

[9] The shift in meaning in what counted as a 'planet' between Ptolemy and Copernicus is similar. We may also compare the points I made before, in chapters 1 and 4, about the relations between 'fire', '*pur*', and '*huo*'. There is no neutral vocabulary to mediate between fire as substance and fire as process, nor between elements in general and 'phases', where I remarked on the tentative moves made by Theophrastus to replace a concept of fire as simple body with one of fire as process (ch. 1 p. 23 n. 31). But the absence of a neutral vocabulary does not mean that all understanding is blocked.

[10] Replacing the literal/metaphorical dichotomy, as I recommend, with an analysis in terms of semantic stretch does not, as I have insisted, threaten the intelligibility of

do the job of interpretation adequately, and of course impossible to do so perfectly.

But we should now consider other cases where incommensurabilities have been diagnosed, including in relation to apparently less obviously theory-laden terms. Let us return to the problems raised by Viveiros de Castro's perspectivism that I mentioned in chapter 1, where jaguars are said to see blood as manioc beer, vultures the maggots in rotting meat as grilled fish, while humans see the blood and maggots as blood and maggots. What the jaguar sees depends on its body, and likewise for vultures, humans, and every living creature. Conversely, while bodies differ radically, all living beings share in social behaviour, again a notion that conflicts with our usual, though evidently challengeable, assumption that nature is the same, while cultures are limited to humans.

Interestingly in two of his discussions, Viveiros de Castro (2004, 2009: 40ff.) cites the sense/reference distinction precisely to show how our normal expectations in this regard are defeated. In the case of the planet Venus, he acknowledges that we treat the referent as the same, but hold that the senses of Morning Star and Evening Star differ. In perspectivism, by contrast, it is a mistake to look for a single referent ('beer' being whatever is tasty, nutritious, and mildly intoxicating), for not only do the referents change, but the senses of 'beer', 'blood', and the rest differ across the 'languages' of different species, exhibiting what Viveiros de Castro labels 'controlled equivocation'. Insofar as the equivocation is 'controlled', one might venture an alternative analysis in terms of exceptional semantic stretch.[11]

---

communication, but allows for its complexity and open-endedness, just as my notion of the multidimensionality of reality draws attention to the different aspects of the objects that we may assume legitimately, that is, non-arbitrarily, to be there to be communicated about. But in neither instance can we specify, in advance, how open-ended the communication may be allowed to be, nor what multidimensionality we are dealing with.

[11] What would a semantic stretch analysis look like? I used that notion to block the application of the literal versus metaphorical dichotomy, when it is demanded that a term be univocal, with a single determinate sense, or else be condemned as merely figurative, or worse, hopelessly equivocal. In the perspectivism case, resistance to the demand for univocity is just as essential, indeed even more so, with regard to the perspectivists' reports on what different kinds of creatures perceive. Jaguars do not drink metaphorical beer: but no more do they drink literal beer either. 'Beer', or rather whatever indigenous term that translates, should be permitted the stretch that I first illustrated with 'green' in Marvell in the last chapter, indeed quite exceptional stretch here that must allow for the central point of perspectivism, that different perceivers each have their own perceptions. I shall shortly discuss, in my text, the more familiar case where, in the Mass, what is wine

But leaving aside the question of how the Achuar and the Araweté—and their shamans[12]—arrive at the ideas they have about jaguars, vultures, anacondas, and how humans and non-humans relate to one another, we may certainly register how radically their views differ from those we generally assume. Faced with this challenge, it may look as if we have to say that we are dealing with a multiplicity of worlds—not just those of the Araweté and jaguars, but also those of the Araweté and ours—between which no communication is possible. But that, I suggest, would be premature, for three reasons.

First there is the general point that I have made before, that while perspectivism conflicts with many of our ordinary assumptions, it is certainly not the case that it is unintelligible. Equally we may or may not go along with one or other of Descola's fourfold ontologies, but that will not be because we cannot understand them. The idea that animals are persons and share in culture (while their bodies or their natures differ) certainly conflicts with twenty-first-century Western beliefs. But if strange, it is not incomprehensible and no more strange (Viveiros de Castro would be right to insist) than much Western metaphysics. Evidently the moral implications of monoculturalism and multinaturalism, in particular, pose a challenge, but however we respond to that, they are not beyond our grasp. Indeed they would hardly pose the kind of challenge they do, if they were simply incomprehensible.

---

to the outsider is the blood of Christ to the faithful, where 'outsider' and 'faithful' differ in their perceptions as much as jaguars and humans do in Amazonia. That shares with perspectivism and the other radical meaning shifts that I have discussed ('force', 'weight', 'mass', 'fire', '*pur*', and '*huo*') that we should resist concluding that only one view of the matter can be correct. However, the faithful who originally saw wine in the chalice (before the consecration) believe that it is the wine that has been changed thanks to the miracle of transubstantiation, rather than that they, the perceivers, have been transformed and now have the vision of another life form.

[12] How much of a gap there may be between shamans and the rest of the collectivities to which they belong is hard to determine in some ethnographic reports. The issue revolves around how someone becomes a shaman, and indeed whether anyone can become one and acquire the powers that are their prime attributes, those of healing, divination, and especially the ability to communicate with spirits and the dead. Sometimes shamans are visionaries whose gifts set them completely apart from ordinary members of the societies to which they belong: but in some accounts they just possess to an exceptional degree talents that are widespread in those societies. For a first orientation in the extensive literature, the studies of Vitebsky 1993, Humphrey and Onon 1996, Hugh-Jones 1996, Viveiros de Castro 2009, and Vilaça 2010 are particularly instructive.

Secondly, while ethnographers may sometimes be puzzled as to the exact commitments their informants make in their statements about jaguars, the fact that those statements are indeed about jaguars is not in doubt. These jaguars are said to see what they see because they have the bodies they have: but the characteristics that make them *jaguars'* bodies are agreed. It is true that a jaguar may be a shaman in disguise,[13] simply wearing the 'clothes' of a jaguar, as in some of the ethnographic reports, and it may take another shaman to see them when they go off and join jaguar society and behave like human beings. But those disguises or clothes can be described and identified as indeed those of jaguars.

Thirdly, and more generally, the very success with which many ethnographers are able to get to grips with and eventually describe the ontological presuppositions, the cosmologies, and the values of the peoples they study shows that some understanding (however incomplete) is within reach, even when those ontologies differ radically from the ones the ethnographers initially entertained. Clearly the vocabularies in question do not readily map onto one another. Yet given that ethnographers provide us with the makings of a translation manual, that should give us pause before we conclude that we are faced with irresoluble incommensurabilities.

Before I leave the perspectivist problem, let me add a further example that will help to clarify the issue and certainly serve to block any notion that the difficulties arise solely from exotic ethnographic reports or antiquated science. In our own collectivity, in modern England, there are devout Roman Catholics who believe that in the Mass the wafer is transformed into the flesh, the wine into the blood, of Christ. Does not the Gospel according to St Matthew, chapter 26, report Jesus saying at the Last Supper that the bread 'is my body' and the wine 'is my blood of the new testament, which is shed for many for the remission of sins'? The transubstantiation case differs from Amerindian perspectivism in that it is the objects—the wafer and the wine— that change, not the perceiving agents. The celebrants remain the same, but one minute they have a wafer in front of them, the next, after the consecration, it is the body of Christ, not just a representation or a symbol of it, but His body. But the point the two types of case have in

---

[13] Conversely a shaman may be a jaguar in disguise.

common is that in the transubstantiation experience too what seems like a wafer to some (the unbelievers) is no wafer but the body of Christ for others (the faithful).

Asked to explain their commitments, believers would no doubt give different responses. Some accept transubstantiation simply on the authority of the Bible and the Church and there is no more to it. It is adherence to the faith that is important, just as in most societies the authority of those who transmit received wisdom is the key to its acceptance. Some Christians will go further and insist that the sacrament is a mystery and glory in its paradoxicality, which they would not be able to do, of course, but for a clear idea of what a paradox is. Others, however, will use the various exegetic devices I outlined in chapter 4 to palliate the apparent counter-intuitiveness of the belief. But the important point for my discussion here is that these believers are members of the same 'advanced' Western society as ourselves and speak our language where there is no possible issue of having to 'translate' in the sense in which that applies to another tongue. They would undoubtedly claim that their understanding of the world, and of the place of humans in it, is special—thanks to their faith they have access to a true vision of what is worthwhile—but they would normally not commit themselves to the idea that they inhabit a radically different world. We are faced with a problem of interpretation that requires all our imagination to begin to resolve. We need once more to take a leaf out of Zhuangzi's book and follow his advice about 'lodge sayings'. We should put ourselves in others' positions, and if they do the same, there is some hope for progress even if much may remain unclear both about the nature of a belief and about the reasons for holding it, and even if criticisms, objections, and differences of opinion do not all disappear.

So when the argument is (as it sometimes is) that different conceptual frameworks may be such that there is no possible communication between them, then the hypothesis is *too* strong. It is possible to build up an interpretation both of Galileo's use of weight and force, and of Aristotle's: commentators are constantly doing just that, with, it is true, greater or less success. Perspectivism poses just as great a challenge, but successful communication between ethnographer and informant even in the matter of ontologies is strong evidence against any conclusion to the effect that mutual incomprehensibility reigns across the board. If there were a human society such that its language was totally unintelligible to outsiders, well, it would be unintelligible: whereof one cannot speak thereof one must remain silent, as Wittgenstein said. But as I have said before, no such society is imaginable.

Both history of science and ethnography give us access to widely diverging systems of beliefs and practices. That is their challenge, but also their promise. Provided we do not accept—as we should surely not accept—that we are the prisoners of our own given conceptual categories and inherited cultural norms, we have a great deal to learn from others, from their thoughts and beliefs and values, their ways of organizing society, their ideas on what is worthwhile in human experience, their skills, their modes of being in the world. The pay-off is thus not just intellectual, but relevant to how we should live, certainly to our own self-understanding. We may never be faced with the challenges of subsistence in the Amazonian jungle nor having to navigate the Pacific by stars and winds and waves alone. But we may reflect with due modesty that the skills we happen to pride ourselves on in our society are not the only ones that humans have brought to a high level of perfection.

A measure of objectivity in evaluation is attainable, even though it is difficult to attain. We should not be hamstrung by the standard Western philosophical dichotomies that have proved such an obstacle to understanding in the past. There are occasions when plural answers to an apparently simple single question are appropriate, even indispensable. That applies to some of those philosophical dichotomies as well as to substantive issues in the understanding of a wide range of phenomena. Such a question as the main genera and species of animals is certainly open-ended, as also is to some extent even the question of what an animal is. For ordinary purposes we assume we know: but when we get into the biology, we find that much of our naïve confidence is misplaced.[14]

Again, it is foolish to attempt to prescribe in advance *how* we should go about investigating the phenomena that interest us whether as scientists or just as ordinary people, that is, the style of inquiry we should adopt, for different styles are far more often complementary

---

[14] Once again, this is not just a phenomenon of modern times. Aristotle found his views on the distinction between animals and plants (the former defined as having perception, the latter as lacking it) were hard to apply in some cases, as I pointed out in Lloyd 1996b: ch. 3.

than mutually exclusive. It is especially short-sighted to deny that anyone else has anything of value to teach us, to suppose that our own initial ontological assumptions (whatever they may be) are the right ones, as if we know everything there is to be known about what there is, about human cognition, and about what should be matters of concern for us as humans. Plenty of thoughts on that subject have been dead-ends, plenty have been not just wrong, but wrong-headed. But that makes it all the more important to make sure that we have done the fullest possible job of first trying to understand. The puzzles are there to challenge us. But if there is much that at the end of our inquiry we cannot accept or condone, there is much more that is suggestive and that can open up perspectives that were initially beyond our imagination.

# Epilogue

I have surveyed a considerable range of material concerning the conceptual systems of a variety of ancient and modern societies, where the questions of why individuals and groups believed what they did are often problematic in the extreme. We have every reason to be puzzled by the content of their reported understandings, and we may also be taken aback at their very variety. On the one hand, human adaptability to new practical circumstances clearly carried an evolutionary advantage, not least in the migration of human populations—and first in the movement out of Africa—where new ecological conditions had to be confronted. On the other, the ability to come up with new solutions to aspects of the human predicament, from social arrangements to making sense of the world around us, is far from limited to circumstances in which survival itself depended on that talent.

In my first chapter, on understandings of what it is to be human, I argued that cosmological and ontological inventiveness is one of the main distinctive characteristics exhibited by humans across time and space—the phenomenon that generates the set of problems that I have been tackling in these studies. The challenge is to come to terms with what the evidence available to us, from ancient and from modern societies, reveals about human imaginative resourcefulness. Some people's experiences and practices evidently have very little in common with others', as do their beliefs about the world they live in. Their forms of life, including their forms of livelihood and their relations with others in their own group and yet others perceived as belonging to other groups, may differ widely. We are used to appreciating heterogeneity and pluralism in music and art, for instance. When we are faced with them also in cosmology and ontology, the leit-motif of my discussion is that this is an opportunity, not a threat.

Admittedly the opportunity to learn from others is sometimes not only not welcomed, but positively rejected and denied. The last thing in the world that some people want to do is to take on board what others, seen as strangers, foreigners, enemies, even as sub-human, think, or the way they behave. Implicitly or explicitly group solidarity is made the cardinal principle, and one can see that for groups to survive, solidarity is often vital. Any lessening, on anyone's part, of their adherence to the group's norms and values will then be construed as a betrayal and the culprit disciplined, ostracized, condemned. Such a reaction may apply not just where the deviant ideas or practices relate to moral matters—to sexual behaviour, for instance—but also where beliefs about physical phenomena are concerned. One ancient Greek philosopher (Anaxagoras) got into trouble because he held that the sun is a red-hot stone not much larger than the Peloponnese. Galileo more famously was punished for believing that the earth moved round the sun. The stronger the institutions that take upon themselves the responsibility for laying down what is to be believed and how people should behave, the more likely it is that any deviant will be penalized for stepping out of line. What may start out as conducive to survival may become a straitjacket inculcating exclusive beliefs that deny the freedom of thought, even the right to be considered human, to much of the human race.

We pride ourselves, no doubt, on the pluralism and tolerance of our own society. But they have their limits. We have our authoritarian institutions, including in our universities and academies, and even in our research laboratories, that seek to impose criteria of acceptability. In plenty of 'advanced' societies, religion is still a major force and sometimes one that is intolerant of the views of other faiths or of lay people generally. The way in which others' ideas and practices have been, and continue to be, dismissed, including by philosophers and anthropologists, as deluded or irrational, should give us pause. I have problematized three major sets of issues in the course of this book, 'being' or ontologies, humanity and what it is to be human, and understanding—the world and each other. It is time to attempt, in conclusion, to review what it seems possible to claim we have learnt from our study of each of the three.

'Being' has repeatedly been treated, in Western philosophy at least, as a problem to be solved by clear-headed application of epistemological criteria. We perceive what there is, and we reason our way to conclusions about that, indeed not just about what is, but about what must be, because reason tells us so. Already a bewildering variety of solutions was suggested in Greek antiquity, and the history of subsequent European philosophy has been dominated by the ongoing but unresolved debates between increasingly sophisticated epistemological proposals, empiricist, rationalist, idealist, phenomenological, realist, constructionist, some of which were discussed in chapter 5. Two recurrent features of Western speculation have been the demand for certainty and the associated requirement that the objects concerning which that certainty is to be achieved must themselves be stable essences.

Yet other traditions of philosophy, in China for instance, have been far less preoccupied with the issue of epistemological foundations and with the quest for incontrovertibility. Meanwhile, the reported cosmological and ontological ideas in the ethnographic literature suggest that in many societies, ideas about bodies, selves, agents, and the practices that correspond to those ideas and are evidence for them, often diverge dramatically from those with which the West has generally been familiar. The data marshalled by Descola, Viveiros de Castro, and others make the point abundantly clear, however, we then set about their classification and analysis. Thus in contrast to an ontology of fixed essences, perspectivism introduces us to one of ever-changing relations, where the relations themselves are never independent of a point of view from which they are apprehended. Evidently this is a world in which the focus on becoming in no way calls for a realm of being that is immune to change, although the very idea that being is to be separated off from becoming represents a prejudicial Platonic way of presenting the issue.

We are all capable of making mistakes, in small matters and in big. But the dismissive attitudes I mentioned not only underestimate the intelligence of those who hold the challenging ideas we have been discussing and who live by their implications. They tend also to oversimplify the nature of ontologies themselves. They are evidently not matters on which simple judgements of truth and falsehood are possible. The question of what there is may have a deceptively straightforward grammatical form: but no single simple answer is going to prove at all satisfactory. In Lloyd (2007) I already suggested with regard to colour perception and spatial cognition in particular that the phenomena are multidimensional and that different styles of investigating them may yield different results that should be treated as complementary rather than alternative. Here I extend those notions in relation to the overall question of being itself. We have every reason to take seriously divergent views. That does not mean that we can agree with and endorse all the ideas and practices we encounter—that 'anything goes' in that sense. But it does mean that we can and should learn not just from the answers to the question of what there is, but also from the associated values, ideals, aspirations, and forms of life of those who have given or implied those answers. True, there is evidence to take into account and arguments to assess, although some ontologies are much more explicit about their bases than others. But we should be wary of assuming that credibility depends on a purpose-built epistemology. That again is a prejudicial Western manner of approaching the issue.

Moreover, there is, of course, no reason to suppose that there is no more to be discovered, in time, from science, from ethnography, and even from ancient history. Where science is concerned, it is not just a matter of discovery, but also one of invention, of bringing new realities into existence, although the distinction between discovery and invention may, in practice, be hard to draw. As all those disciplines expand, science, of course, especially the scope of the ontological agenda, will grow correspondingly. That it will grow is clear: how it will, no one is in a position to say. But rather than expect some eventual resolution of the issue between ontologies of being and ontologies of becoming, for example, we should aim, in my view, for a fuller understanding of how they may be complementary.

That takes me to my second theme of humanity, where, precisely, those other factors influencing ontologies and indeed constitutive of them can be discerned, where values, ideas about what makes life worth living, a view of the human condition, and what defines that in contrast to whatever is perceived as non-human, are generally deeply implicated. Here too an amazing proliferation of beliefs and practices is attested across the world and throughout human history, and they have as often been exclusive—refusing humanity to what biology or mere appearance would label other humans—as inclusive or universalist.

As I argued in chapter 1, simple solutions to the problem of the defining characteristics of what makes humans human tend to mislead. What links all human beings is not just a certain basic anatomy and physiology and a shared genetic inheritance, but also the fact that we all undergo processes of social acculturation. Conversely what differentiates us from one another is not just the actual cultures we acquire, but also certain features of our biology, our susceptibility to certain diseases, for instance, and even our varying abilities to discriminate colours. 'Nature' and 'culture', as traditionally conceived, do not do an at all adequate job of discriminating what is common and what is not across human populations, and that is before we factor in that those human populations have entertained such different views on the answer to precisely that question.

An encounter with a very different set of moral norms poses a particular challenge. We may end up by rejecting them for ourselves, but that cannot be where we should begin. Rather, as with ontologies in general, the first task is to investigate how those norms operate, what may have stimulated their development, what factors contribute to their continued maintenance, in short, to understand them. But again, as with ontologies, the very fact that such diverse proposals and practices have been developed by different human individuals and groups across space and time tells us something important about humanity, and its, our, flexibility and imagination, and indeed also our fallibility, our credulity.

We have a marvellous gift for creating cosmologies, with different ideas about our, human, place in the complexity of things. But that talent may be put to perverse use when a sense of our own exclusive rightness leads to the denial of anyone else's claim to a hearing, even a claim to be human. As we have seen, the tendency to restrict who is to count as a proper human to the members of the group to which the speaker happens to belong is widespread, and once such a conviction is entrenched, it takes particular determination and courage to uproot it. Human inventiveness is far from always directed to enriching our sense of what we as humans all share, for it is all too commonly used to show how very much superior 'we' are to everyone else. To understand is, as I have emphasized, not to condone, but not to attempt to understand is to deny that we all share that gift. Besides, we learn from understanding even what indeed we do not ultimately condone, and that may include revising our ideas about whether to condone. Comparative ontology, as I practise it, puts us on the spot. There is no algorithm to determine when we should welcome, or when we should be wary of, systems of belief that differ radically from our own. But the very endeavour of trying to comprehend them affords us a notable opportunity to reflect critically on our own assumptions and values, a capacity that we certainly need urgently to develop, given our current geopolitical and environmental predicaments and the dangers posed on the one hand by anarchic materialism, on the other by ideological dogmatism. Those traits have, of course, existed before, but never previously combined with the powers of appropriation and destruction that are now within reach.

But if understanding, of self, of others, and of the world, is a generally accepted goal, very varying opinions have been entertained on how the goal itself is to be construed. A traditional view had it that understanding is a matter of describing and explaining how things really are, on the model of what the historian von Ranke claimed should be the aim of history, namely to recount what really happened, *wie es eigentlich gewesen*. What is there to be understood, on that view, is given independently of our understanding, and our understanding is simply a matter of getting it right. But we have seen reason to doubt those assumptions. Already attempting to understand other people raises problems, insofar as we discover that their understanding of the world and of themselves often differs radically from ours. That does not mean that all claimed understandings are justified, all are correct, and there is no such thing as misunderstanding—the problem I tackled in the chapter on error. But it does mean that we should not assume at the outset that there is just one simple truth of the matter to be had, let alone one to which we happen to have exclusive access ourselves. Thus what it is to understand is not (we have learnt) just a matter of giving causal accounts, let alone ones presented as the conclusions to deductive arguments, though that is an ideal that has reverberated in Western thought ever since Aristotle. Often in other traditions understanding turns on being able to trace the associations between things, their correlations, or again on putting yourself in someone else's position, to see matters from their point of view.

A recurrent problem relates to how far it is possible to stand outside one's inherited assumptions, norms, beliefs, to view them critically, even if that is permitted to us in the collectivity in which we live and it is what we sense we should do. Access to others' systems of beliefs—an experience not confined to anthropology and to history as academic disciplines, of course—may raise the question, though it is one that may (as I have said) be immediately foreclosed by a curt dismissal of the validity of those beliefs. But even when we take others' initially puzzling views and behaviour seriously, how do we avoid mere self-deception when we claim to have grasped their meaning? There is, of course, no guarantee of success. The exploration is never complete and such results as may be suggested must always be treated as provisional and revisable. Those remarks apply equally to this study. There is, as I said, for sure, more to be learnt in due time, but this discussion is offered as a contribution to a study of how the questions may be addressed under three main heads, concerning the principles that should guide our investigation of ontologies.

The first is the reminder that styles of inquiry vary and when they do may be complementary, not alternative. The second is that when, as often, the phenomena investigated are multidimensional, it is again neither necessary nor desirable to plump for one view to the exclusion of all others. My third positive, if controversial, proposal relates to the idea of semantic stretch, which allows the exploration of divergent senses of individual terms and collocations without the restrictions imposed by the literal/metaphorical dichotomy. That notion, combined with those two ideas of the recognition of the multidimensionality of many phenomena and the plurality of possible styles of inquiry, provides, I believe, the best toolkit to approach the understanding of complex ontologies. With semantic stretch in play the investigation is more open-ended and deploys more of the resources associated with the interpretation of poetry than scientists or philosophers normally accept. This, I believe, is what is needed if we are to do justice to that imaginative resourcefulness that I have spoken about. But that is certainly not to make understanding any easier, but to redefine the nature of the difficulties involved. It is, however, to insist that we should replace the desire for simple answers with a riskier acceptance of the provisionality and conditionality of any grasp of the issues.

To entertain and explore radically different views of reality, of humanity, and of attempts to understand is hard work, challenging, and uncomfortable. But the study of others' ideas and practices, undertaken in that spirit of tentative investigation, has a lot to offer, for we learn both about the diversity of lifestyles and of the fundamental beliefs that have been held by different individuals and groups across the world, and about the circumstances in which, within a single society, divergence on those issues is allowed and even flourishes. Exploiting that possibility in my own society, I have undertaken this exploration of the questions.

# GLOSSARY OF KEY CHINESE TERMS AND NAMES

| | | |
|---|---|---|
| *an* | 安 | where, on what basis |
| *bienao* | 鱉臑 | pyramid with right triangular base and one lateral edge perpendicular to the base |
| *bu ran* | 不然 | not so |
| *dao* | 道 | the way |
| *Er ya* | 爾雅 | (third century BCE encyclopaedia) |
| *fa jia* | 法家 | 'School of Law' |
| *fei* | 非 | no, is not |
| Gaozi | 告子 | (philosopher of fourth century BCE) |
| Gongsun Long | 公孫龍 | (philosopher of 'School of Names') |
| Han Fei | 韓非 | (philosopher of 'School of Law') |
| Han Wu Di | 漢武帝 | ('Martial Emperor' of Han dynasty) |
| *he yi* | 何以 | by what means |
| *Hong Fan* | 洪範 | 'Great Plan' section of *Shangshu*, 尚書 |
| Hou Ji | 后稷 | (Lord of Millet) |
| *Huainanzi* | 淮南子 | (second-century BCE cosmological text) |
| *Huangdi neijing* | 黃帝內經 | *Inner canon of the Yellow Emperor* |
| Hui Shi | 惠施 | (philosopher of 'School of Names') |
| *huo* | 火 | fire, 'flaming upwards' |
| Jian Di | 簡狄 | (Legendary founder of Yin dynasty) |
| Jiang Yuan | 姜原 | (Mother of Hou Ji) |
| *jin* | 金 | metal |
| *Jiuzhang suanshu* | 九章算術 | *Nine Chapters on Mathematical Procedures* |
| *junzi* | 君子 | 'gentleman' |

| Kong Fuzi | 孔夫子 | Confucius, purported author of *Lun Yu* 論語 (*Analects*) |
| *lifa* | 曆法 | 'calendar studies' |
| Liu An | 劉安 | (king of Huainan) |
| Liu Hui | 劉徽 | (third-century CE mathematician) |
| Lü Buwei | 呂不韋 | (compiler of *Lüshi chunqiu*) |
| *luan* | 亂 | disorder, chaos, anarchy |
| *Lüshi chunqiu* | 呂氏春秋 | (third-century BCE cosmological text) |
| Mengzi | 孟子 | Mencius |
| Mozi | 墨子 | (philosopher, founder of Mohists) |
| *qi* | 氣 | breath/energy |
| Qin Shi Huang Di | 秦始皇帝 | (first emperor) |
| *ran* | 然 | so |
| *ren* | 人 | human |
| Shen Gua | 沈括 | (eleventh-century polymath) |
| *shen hua* | 神話 | 'spirit talk', myth |
| *sheng* | 生 | life |
| *shi* | 是 | yes, is the case |
| *Shiji* | 史記 | (first of the dynastic histories) |
| *shu shu* | 數術 | 'calculations and methods' |
| *shui* | 水 | water, 'soaking downwards' |
| Sima Qian | 司馬遷 | (co-author of *Shiji*) |
| *suan* | 算 | reckoning |
| *suan shu* | 算術 | art of reckoning, mathematics |
| *Suanshushu* | 算數書 | (second-century BCE mathematical text) |
| *tian* | 天 | heaven |
| *tian wen* | 天文 | patterns in the heavens |
| *wang* | 王 | king |
| Wang Chong | 王充 | (first-century philosopher, author of *Lun Heng* 論衡) |
| *wu xing* | 五行 | five phases |
| Xunzi | 荀子 | (third-century BCE philosopher) |
| *yi* | 義 | right, righteousness |
| *yin shi* | 因是 | rely on |
| *yin yang* | 陰陽 | negative and positive principles |
| *youshui* | 遊說 | 'wandering advisers' |

| | | |
|---|---|---|
| *yu yan* | 寓言 | 'lodge sayings' |
| Zhao Youqin | 趙友欽 | (thirteenth-century mathematician) |
| *zheng ming* | 正名 | rectification of names |
| *zhi* | 知 | knowledge |
| *zhi yan* | 巵言 | 'spill-over sayings' |
| *zhong yan* | 重言 | 'weighty sayings' |
| *Zhoubi suanjing* | 周髀算經 | *Arithmetic Classic of the Zhou Gnomon* |
| Zhuangzi | 莊子 | (fourth-century BCE philosopher) |

# BIBLIOGRAPHY

AGAMBEN, G. (2004) *The Open: man and animal* (Stanford)

ATRAN, S. (2001) 'The case for modularity: sin or salvation?' *Evolution and Cognition* 7: 46–55

——and MEDIN, D. (2008) *The Native Mind and the Cultural Construction of Nature* (Cambridge, Mass.)

AVITAL, E. and JABLONKA, E. (2000) *Animal Traditions: Behavioural Inheritance in Evolution* (Cambridge)

BARKER, A.D. (2007) *The Science of Harmonics in Classical Greece* (Cambridge)

BARKOW, J.H., COSMIDES, L., and TOOBY, J. (eds.) (1992) *The Adapted Mind: Evolutionary Psychology and the Generation of Culture* (Oxford)

BARTH, F. (1975) *Ritual and Knowledge among the Baktaman of New Guinea* (Oslo)

BATESON, P. and MAMELI, M. (2007) 'The innate and the acquired: useful clusters or a residual distinction from folk biology?' *Developmental Psychology* 49: 818–31

BEARDSMORE, R.W. (1996) 'If a lion could talk', in K.S. Johannesen and T. Nordenstam (eds.) *Wittgenstein and the Philosophy of Culture* (Vienna), 41–59

BLACK, M. (1962) *Models and Metaphors* (Ithaca, NY)

BOESCH, C. (1996) 'The emergence of cultures among wild chimpanzees', in Runciman, Maynard Smith and Dunbar (1996), 251–68

——and BOESCH, H. (1984) 'Possible causes of sex differences in the use of natural hammers by wild chimpanzees', *Journal of Human Evolution* 13: 415–40

BOYD, R. and RICHERSON, P.J. (2005) *The Origin and Evolution of Cultures* (Oxford)

BOYER, P. (1994) *The Naturalness of Religious Ideas* (Berkeley)

——(2001) *Religion Explained* (London)

——and BARRETT, H.C. (2005) 'Domain specificity and intuitive ontology', in D.M. Buss (ed.) *The Handbook of Evolutionary Psychology* (Hoboken), 96–118

BROWN, D. (2000) *Mesopotamian Planetary Astronomy–Astrology* (Groningen)

BURCKHARDT, J. (1898–1902) *Griechische Kulturgeschichte*, 2nd edn., 4 vols. (Berlin)

BYRNE, R. and WHITEN, A. (eds.) (1988) *Machiavellian Intelligence: Social Expertise and the Evolution of Intellect in Monkeys, Apes, and Humans* (Oxford)

CAREY, S. (2009) *The Origin of Concepts* (Oxford)

CARRITHERS, M., COLLINS, S., and LUKES, S. (1985) *The Category of the Person* (Cambridge)

CARRUTHERS, P. and CHAMBERLAIN, A. (2000) *Evolution and the Human Mind* (Cambridge)

CATCHPOLE, C.K. and SLATER, P.J.B. (2008) *Bird Song*, 2nd ed. (1st edn. 1995) (Cambridge)

CHANG, H. (2012) *Is Water H$_2$O? Evidence, Realism and Pluralism* (Dordrecht)

CHEMLA, K. (2003) 'Generality above Abstraction: the General Expressed in Terms of the Paradigmatic in Mathematics in Ancient China', *Science in Context* 16: 413–58

——and GUO SHUCHUN (2004) *Les Neuf chapitres: Le Classique mathématique de la Chine ancienne et ses commentaires* (Paris)

CHEN, CHENG-YIH (1987) 'The Generation of Chromatic Scales in the Chinese Bronze Set-Bells of the −5th century', in Chen, Cheng-Yih et al. (eds.) *Science and Technology in Chinese Civilization* (Singapore), 155–97

CHENEY, D.L. and SEYFARTH, R.M. (1990) *How monkeys see the world* (Chicago)

CHENG, A. (ed.) (2005) 'Y a-t-il une philosophie chinoise? Un état de la question', *Extrême-Orient Extrême-Occident* 27

CLARK, S.R.L. (1997) *Animals and their moral standing* (London)

COLLINS, R. (1998) *The Sociology of Philosophies* (Cambridge, Mass.)

COSMIDES, L. and TOOBY, J. (1992) 'Cognitive Adaptations for Social Exchange', in Barkow, Cosmides, and Tooby (1992), 163–228

CROMBIE, A.C. (1994) *Styles of Scientific Thinking in the European Tradition*, 3 vols. (London)

CULLEN, C. (1996) *Astronomy and Mathematics in Ancient China: the Zhou bi suan jing* (London)

——(2000) 'Seeing the Appearances: Ecliptic and Equator in the Eastern Han', *Studies in the History of Natural Sciences* 19: 352–82

——(2004) *The Suan Shu Shu: Writings on Reckoning* (Needham Research Institute Working Papers 1) (Cambridge)

——(2007) 'Actors, networks and "disturbing spectacles" in institutional science: 2nd century Chinese debates on astronomy', *Antiquorum Philosophia* 1: 237–67

CUOMO, S. (2001) *Ancient Mathematics* (London)

DARWIN, C. (1859) *The Origin of Species* (London)

DASCAL, M. (2006) *Gottfried Wilhelm Leibniz: The Art of Controversies* (Dordrecht)

DASTON, L. and GALISON, P. (2007) *Objectivity* (New York)

DAVIDSON, D. (2001) *Essays on Actions and Events*, 2nd edn. (Oxford)

DELPLA, I. (2001) *Quine, Davidson. Le principe de charité* (Paris)

DESCOLA, P. (1992) 'Societies of nature and the nature of society', in A. Kuper (ed.) *Conceptualizing Society* (London), 107–26

——(1996) 'Constructing natures: symbolic ecology and social practice', in P. Descola and G. Pálsson (eds.) *Nature and Society: Anthropological Perspectives* (London), 82–102

——(2005) *Par-delà nature et culture* (Paris)

——(2009) 'Human natures', *Social Anthropology* 17: 145–57

DETIENNE, M. (1967/1996) *The Masters of Truth in Archaic Greece* (trans. J. Lloyd of *Les Maîtres de vérité dans la Grèce archaïque* (Paris, 1967)) (New York)

——(1972) 'Entre bêtes et dieux', *Nouvelle revue de psychanalyse* 6: 231–46 (revised and retitled version reprinted in Detienne 1977/1979)

——(1972/1977) *The Gardens of Adonis* (trans. J. Lloyd of *Les Jardins d'Adonis* (Paris, 1972)) (Hassocks)

——(1977/1979) *Dionysos Slain* (trans. M. and L. Muellner of *Dionysos mis à mort* (Paris, 1977)) (Baltimore)

——(ed.) (2003) *Qui veut prendre la parole?* (Paris)

——and VERNANT, J.-P. (1978) *Cunning Intelligence in Greek Culture and Society* (trans. J. Lloyd of *Les Ruses de l'intelligence: la mètis des Grecs* (Paris, 1974)) (Hassocks)

DUNBAR, R.I.M. (1988) *Primate social systems* (London)

——(1999) 'Culture, honesty and the freerider problem', in R.I.M. Dunbar, C. Knight, and C. Power (eds.) *The Evolution of Culture* (Edinburgh), 194–213

——(2008) 'Mind the Gap: Why Humans are not just Great Apes', *Proceedings of the British Academy* 154: 403–23

DUPRÉ, J. (2002) *Humans and Other Animals* (Oxford)

ELLEN, R. and FUKUI, K. (eds.) (1996) *Redefining Nature* (Oxford)

ENFIELD, N.J. and LEVINSON, S.C. (eds.) (2006) *Roots of Human Sociality: Culture, Cognition and Interaction* (Oxford)

EVANS-PRITCHARD, E.E. (1956) *Nuer Religion* (Oxford)

FEYERABEND, P.K. (1975) *Against Method* (London)

FODOR, J. (1983) *The Modularity of Mind* (Cambridge, Mass.)

FOUCAULT, M. (1965) *Madness and Civilization* (trans. R. Howard of *Histoire de la folie* (Paris, 1961)) (New York)

GARDNER, R.A., GARDNER, B.T., and van CANTFORT, T.E. (eds.) (1989) *Teaching sign language to chimpanzees* (Albany, NY)

GASSMANN, R.H. (1988) *Cheng Ming. Richtigstellung der Bezeichnungen. Zu den Quellen eines Philosophems im antiken China. Ein Beitrag zur Konfuzius-Forschung* (Etudes asiatiques suisses 7, Berne)

GEERTZ, C. (1973) *The Interpretation of Cultures* (New York)

GELLNER, E. (1985) *Relativism and the Social Sciences* (Cambridge)

GELMAN, R. and SPELKE, E. (1981) 'The development of thoughts about animate and inanimate objects: implications for research on social cognition', in J.H. Flavell and L. Ross (eds.) *Social Cognitive Development* (Cambridge), 43–66

GERNET, J. (1985) *China and the Christian Impact* (trans. J. Lloyd of *Chine et christianisme* (Paris, 1982)) (Cambridge)

GINZBURG, C. (2002) *Wooden Eyes* (trans. M. Ryle and K. Soper of *Occhiacci di legno* (Milan, 1998)) (London)

GOODMAN, N. (1978) *Ways of Worldmaking* (Hassocks)

GRAHAM, A.C. (1978) *Later Mohist Logic, Ethics and Science* (London)

——(1989) *Disputers of the Tao* (La Salle, Ill.)

GRICE, H.P. (1968) 'Utterer's meaning, sentence-meaning, and word-meaning', *Foundations of Language* 4: 225–42 (reprinted in Grice 1989: 117–37)

——(1975) 'Logic and conversation', in P. Cole and J. Morgan (eds.) *Syntax and Semantics* vol. 3, *Speech Acts* (New York), 41–58 (reprinted in Grice 1989: 22–40)

——(1978) 'Further notes on logic and conversation', in P. Cole (ed.) *Syntax and Semantics* vol. 9, *Pragmatics* (New York), 113–27 (reprinted in Grice 1989: 41–57)

——(1989) *Studies in the Way of Words* (Cambridge, Mass.)

GRIFFIN, D.R. (1974) *Listening in the Dark* (2nd edn.) (New York) (1st ed. 1958)

——(1984) *Animal Thinking* (Cambridge, Mass.)

——(1992) *Animal Minds* (Chicago)

GUTHRIE, W.K.C. (1969) *A History of Greek Philosophy* vol. 3, *The Fifth-Century Enlightenment* (Cambridge)

HACKING, I. (1992) ' "Style" for historians and philosophers', *Studies in History and Philosophy of Science* 23: 1–20

——(2009) *Scientific Reason* (Taipei)

HALLOWELL, A.I. (1960) 'Ojibwa Ontology, Behavior, and World Views', in S. Diamond (ed.) *Culture in History: Essays in Honor of Paul Radin* (New York), 19–52

HARBSMEIER, C. (1998) *Science and Civilisation in China*, vol. VII part 1, *Language and Logic* (Cambridge)

HAUSER, M.D. (1996) *The Evolution of Communication* (Cambridge, Mass.)

HENARE, A., HOLBRAAD, M., and WASTEL, S. (eds.) (2007) *Thinking Through Things* (London)

HESSE, M. (1974) *The Structure of Scientific Inference* (London)

——(1983) 'The Cognitive Claims of Metaphor', in J.P. van Noppen (ed.) *Metaphor and Religion* (Brussels), 27–45

HINDE, R.A. (1999) *Why Gods Persist* (London)

HOWELL, S. (1984) *Society and Cosmos: Chewong of Peninsular Malaysia* (Oxford)

HUANG YILONG (2001) 'Astronomia e astrologia', in S. Petruccioli et al. (eds.) *Storia della Scienza* vol. 2, sez. 1, ch. 13 part 4, 167–70

——and CHANG CHIH-CH'ENG (1996) 'The Evolution and Decline of the Ancient Chinese Practice of Watching for the Ethers', *Chinese Science* 13: 82–106

HUGH-JONES, S. (1996) 'Shamans, prophets, priests and pastors', in N. Thomas and C. Humphrey (eds.) *Shamanism, History and the State* (Ann Arbor), 32–75

HUMLE, T. and MATSUZAWA, T. (2002) 'Ant-dipping among the chimpanzees of Bossou, Guinea, and some comparisons with other sites', *American Journal of Primatology* 58: 133–48

HUMPHREY, C. and ONON, U. (1996) *Shamans and Elders* (Oxford)

HUMPHREY, N. (2006) *Seeing Red* (Cambridge, Mass.)

——(2011) *Soul Dust* (Chicago)

HUNGER, H. (1992) *Astrological Reports to Assyrian Kings* (State Archives of Assyria 8, Helsinki)

——and PINGREE, D. (1999) *Astral Sciences in Mesopotamia* (Leiden)

——and SACHS, A.J. (1988–2006) *Astronomical Diaries and Related Texts from Babylonia*, 6 vols. (Vienna)

HUTCHINS, E. (1980) *Culture and Inference* (Cambridge, Mass.)

INGOLD, T. (1988) 'Introduction', in T. Ingold (ed.) *What is an animal?* (London), 1–16

——(2000) *The Perception of the Environment* (London)

——(2002) 'Humanity and Animality', in T. Ingold (ed.) *Companion Encyclopedia of Anthropology* (2nd edn.) (London), 14–32

——(2008) 'Anthropology is *not* Ethnography', *Proceedings of the British Academy* 154: 69–92

JOHNSTON, I. (2010) *The Mozi: A Complete Translation* (Hong Kong)

KAHNEMAN, D., SLOVIC, P., and TVERSKY, A. (eds.) (1982) *Judgement under Uncertainty: heuristics and biases* (Cambridge)

KARLGREN, B. (1950) 'The Book of Documents', *Bulletin of the Museum of Far Eastern Antiquities* 22: 1–81

KARMILOFF-SMITH, A. (1992) *Beyond Modularity* (Cambridge, Mass.)

KELLER, E.F. (2000) *The Century of the Gene* (Cambridge, Mass.)

KNOBLOCK, J. (1988–94) *Xunzi: A Translation and Study of the Complete Works*, 3 vols. (Stanford)

——and RIEGEL, J. (2000) *The Annals of Lü Buwei* (Stanford)

KROEBER, A.L. and KLUCKHOHN, C. (1952) *Culture: A critical review of concepts and definitions* (Peabody Museum of American Archeology and Ethnology 47) (Cambridge, Mass.)

KUHN, T.S. (1970) *The Structure of Scientific Revolutions* (first publ. 1962) 2nd ed. (Chicago)

KUMMER, H. (1995) 'Causal knowledge in animals', in Sperber, Premack, and Premack (1995), 26–36

KUPER, A. (1999) *Culture: The Anthropologists' Account* (Cambridge, Mass.)

KUSCH, M. (2010) 'Hacking's historical epistemology: a critique of styles of reasoning', *Studies in History and Philosophy of Science* 41: 158–73

LACKNER, M. (1993) 'La Portée des événements – Réflexions néo-confucéennes sur la "rectification des noms" (*Entretiens* 13.3)', *Extrême-Orient Extrême-Occident* 15: 75–87

LATOUR, B. (2009) 'Perspectivism: "Type" or "bomb"?', *Anthropology Today* 25.2: 1–2

LAU, D. (2006) *Metaphertheorien der Antike und ihre philosophischen Prinzipien* (Frankfurt am Main)

LEACH, E.R. (1972) *Humanity and Animality* (London)

——(1982) *Social Anthropology* (London)

LEVINSON, S.C. (1983) *Pragmatics* (Cambridge)

——(2003) *Space in Language and Cognition* (Cambridge)

——and JAISSON, P. (eds.) (2006) *Evolution and Culture* (Cambridge, Mass.)

LÉVI-STRAUSS, C. (1955/1973) *Tristes tropiques* (trans. J. and D. Weightman of *Tristes tropiques* (Paris, 1955)) (London)

——(1958/1968) *Structural Anthropology* (trans. C. Jacobson and B. G. Schoepf of *Anthropologie structurale* (Paris, 1958)) (London)

——(1962/1966) *The Savage Mind* (trans. of *La Pensée sauvage* (Paris 1962)) (London)

——(1964–71/1970–81) *An Introduction to a Science of Mythology* (trans. J. and D. Weightman of *Mythologiques*, 4 vols. (Paris 1964–71)) 4 vols. (London)

——(1973/1976) *Structural Anthropology II* (trans. M. Layton of *Anthropologie structurale deux* (Paris, 1973)) (New York)

LEWIS, M.E. (1999) *Writing and Authority in Early China* (Albany, NY)

LLOYD, G.E.R. (1983) *Science, Folklore, Ideology* (Cambridge)

——(1987) *The Revolutions of Wisdom* (Berkeley)

——(1990) *Demystifying Mentalities* (Cambridge)

——(1991) 'The invention of nature', in *Methods and Problems in Greek Science* (originally Herbert Spencer Lecture, Oxford, 1989) (Cambridge) ch. 18

——(1996a) *Adversaries and Authorities* (Cambridge)

——(1996b) *Aristotelian Explorations* (Cambridge)

——(2002) *The Ambitions of Curiosity* (Cambridge)

——(2003) *In the Grip of Disease* (Oxford)

——(2004) *Ancient Worlds, Modern Reflections* (Oxford)

——(2005) *The Delusions of Invulnerability* (London)

——(2007) *Cognitive Variations* (Oxford)

——(2009) *Disciplines in the Making* (Oxford)

——(2010a) 'The techniques of persuasion and the rhetoric of disorder (*luan*) in late Zhangguo and Western Han texts', in M. Nylan and M.A.N. Loewe (eds.) *China's Early Empires* (Cambridge) ch. 19: 451–60

——(2010b) 'History and Human Nature: Cross-cultural universals and cultural relativities', *Interdisciplinary Science Reviews* 35: 201–14

——(forthcoming) 'Aristotle on the natural sociability, skills and intelligence of animals', in V. Harte and M. Lane (eds.) Politeia *in Greek and Roman Philosophy* (Cambridge)

——and SIVIN, N. (2002) *The Way and the Word* (New Haven)

MAJOR, J.S. (1993) *Heaven and Earth in Early Han Thought* (Albany, NY)

MARLER, P. (1991) 'Differences in behavioural development in closely related species: birdsong', in P. Bateson (ed.) *The Development and Integration of Behaviour* (Cambridge), 41–70

MARTZLOFF, J.-C. (2009) *Le Calendrier chinois: structure et calculs (104 av. J-C. –1644)* (Paris)

MAUSS, M. (1938) 'Une catégorie de l'esprit humain: la notion de personne, celle de "moi". Un plan de travail', *Journal of the Royal Anthropological Institute* 68: 263–81

MAYR, E. (ed.) (1957) *The Species Problem*, American Association for the Advancement of Science Publications 50 (Washington, DC)

MCGREW, W.C. (1992) *Chimpanzee Material Culture* (Cambridge)

MCMULLIN, E. (2009) 'The Galileo Affair: Two Decisions', *Journal of the History of Astronomy* 40.2: 191–212

MOSKO, M. (2010) 'Partible penitents: dividual personhood and Christian practice in Melanesia and the West', *Journal of the Royal Anthropological Institute* 16: 215–40

NAGEL, T. (1974) 'What is it like to be a bat?', *Philosophical Review* 83: 435–50 (repr. in *Mortal Questions* (Cambridge, 1979), 165–80)

——(1986) *The View from Nowhere* (Oxford)

NEEDHAM, J. (1954– ) *Science and Civilisation in China*, 24 vols. to date (Cambridge)

NETTLE, D. (2009) 'Beyond nature versus culture: cultural variation as an evolved characteristic', *Journal of the Royal Anthropological Institute* 15.2: 223–40

——and Dunbar, R.I.M. (1997) 'Social Markers and the Evolution of Reciprocal Exchange', *Current Anthropology* 38.1: 93–9

Netz, R. (1999) *The Shaping of Deduction in Greek Mathematics* (Cambridge)

——(2009) *Ludic Proof* (Cambridge)

Neugebauer, O. (1975) *A History of Ancient Mathematical Astronomy* 3 vols. (Berlin)

Osborne, C. (2007) *Dumb Beasts and Dead Philosophers. Humanity and the Humane in Ancient Philosophy and Literature* (Oxford)

Panksepp, J. and Panksepp, J.B. (2000) 'The seven sins of evolutionary psychology', *Evolution and Cognition* 6: 108–31.

Parpola, S. (1993) *Letters from Assyrian and Babylonian Scholars* (State Archives of Assyria, 10, Helsinki)

Pinker, S. (1994) *The Language Instinct* (London)

Porzig, W. (1934) 'Wesenhafte Bedeutungsbeziehungen', *Beiträge zur Geschichte der deutschen Sprache und Literatur* 58: 70–97

Premack, D. (1976) *Intelligence in ape and man* (Hillsdale, NJ)

——and Premack, A.J. (1983) *The Mind of an Ape* (New York)

Quine, W. van O. (1960) *Word and Object* (Cambridge, Mass.)

Reiner, E. and Pingree, D. (1981) *Babylonian Planetary Omens, part 2: Enūma Anu Enlil Tablets 50–51* (Bibliotheca Mesopotamica 2, 2, Malibu)

——(1998) *Babylonian Planetary Omens, part 3* (Groningen)

Richerson, P.J. and Boyd, R. (2005) *Not by Genes Alone* (Chicago)

Ricoeur, P. (1978) *The Rule of Metaphor* (trans. R. Czerny of *La Métaphore vive* (Paris, 1975)) (London)

Robson, E. (1999) *Mesopotamian Mathematics, 2100–1600 BC* (Oxford)

——(2009) 'Mathematics education in an Old Babylonian scribal school', in E. Robson and J. Stedall (eds.) *The Oxford Handbook of the History of Mathematics* (Oxford), 199–227

Rochberg, F. (2004) *The Heavenly Writing* (Cambridge)

Rolf, E. (2005) *Metaphertheorien: Typologie, Darstellung, Bibliographie* (Berlin)

Runciman, W.G. (2009) *The Theory of Cultural and Social Selection* (Cambridge)

——, Maynard Smith, J., and Dunbar, R.I.M. (eds.) (1996) 'Evolution of Social Behaviour Patterns in Primates and Man', *Proceedings of the British Academy* 88

Sahlins, M. (2008) *The Western Illusion of Human Nature* (Chicago)

Sedley, D.N. (2007) *Creationism and its Critics in Antiquity* (Berkeley)

Seyfarth, R.M. and Cheney, D.L. (1982) 'How monkeys see the world: a review of recent research on East African vervet monkeys', in C.T. Snowdon,

C.H. Brown, and M.R. Petersen (eds.) *Primate Communication* (Cambridge), 239–52

——(1984) 'Grooming, alliances and reciprocal altruism in vervet monkeys', *Nature* 308: 541–3

SHIROKOGOROFF, S.M. (1935) *Psychomental complex of the Tungus* (London)

SINGER, P. (1976) *Animal Liberation* (London)

SIVIN, N. (1995) *Science in ancient China: Researches and reflections* vol. 1 (Aldershot)

SMITH, B. HERRNSTEIN (1997) *Belief and Resistance* (Cambridge, Mass.)

——(2011) 'The Chimera of Relativism', *Common Knowledge* 17, 1: 13–26

SMITH, R. (2007) *Being Human* (Manchester)

SORABJI, R. (1993) *Animal Minds and Human Morals* (London)

SPERBER, D. (1985) *On Anthropological Knowledge* (Cambridge)

——(1994) 'The modularity of thought and the epidemiology of representations', in L.A. Hirschfeld and S.A. Gelman (eds.) *Mapping the Mind: Domain Specificity in Cognition and Culture* (Cambridge), 39–67

——(1996) *Explaining Culture* (Oxford)

——PREMACK, D., and PREMACK, A.J. (eds.) (1995) *Causal Cognition: A multidisciplinary debate* (Oxford)

——and WILSON, D. (1995) *Relevance* 2nd ed. (1st edn. 1986) (Oxford)

STERELNY, K. (2003) *Thought in a Hostile World* (Oxford)

STRATHERN, M. (1988) *The Gender of the Gift* (Berkeley)

——(1991) *Partial Connections* (ASAO Special Publications 3) (Savage, MD)

——(1999) *Property, Substance and Effect: Anthropological Essays on Persons and Things* (London)

——(2005) *Kinship, Law and the Unexpected* (Cambridge)

SWERDLOW, N.M. (1998) *The Babylonian Theory of the Planets* (Princeton)

TAMBIAH, S.J. (1990) *Magic, Science, Religion, and the Scope of Rationality* (Cambridge)

TOMASELLO, M. (1999) *The Cultural Origins of Human Cognition* (Cambridge, Mass.)

TOOBY, J. and COSMIDES, L. (1992) 'The psychological foundations of culture', in Barkow, Cosmides, and Tooby (1992), 19–136

——(1996) 'Friendship and the banker's paradox: Other pathways to the evolution of adaptations for altruism', in Runciman, Maynard Smith, and Dunbar (1996), 119–43

TVERSKY, A. and KAHNEMAN, D. (1974) 'Judgement under uncertainty: heuristics and biases', *Science* 185: 1124–31

TYBJERG, K. (2004) 'Hero of Alexandria's Mechanical Geometry', in P. Lang (ed.) *Re-Inventions: Essays on Hellenistic and Early Roman Science* (Kelowna), 29–56

VERNANT, J.-P. (1980) 'Between the Beasts and the Gods' (originally published as Introduction to M. Detienne 1972/1977) reprinted in *Myth and Society in Ancient Greece*, trans. J. Lloyd (Hassocks), 130–67

VIDAL-NAQUET (1975) 'Bêtes, hommes et dieux chez les Grecs', in L. Poliakov (ed.) *Hommes et bêtes: entretiens sur le racisme* (Paris), 129–42

VILAÇA, A. (2009) 'Conversion, predation and perspective', in A. Vilaça and R. Wright (eds.) (2009), 147–66

——(2010) *Strange Enemies: Indigenous Agency and Scenes of Encounters in Amazonia* (trans. D. Rodgers of *Quem somos nós: Os Wari' encontram os brancos* (Rio de Janeiro 2006)) (Durham, NC)

——(2011) 'Dividuality in Amazonia: God, the Devil, and the constitution of personhood in Wari' Christianity', *Journal of the Royal Anthropological Institute* 17.2: 243–62

——and WRIGHT, R. (eds.) (2009) *Native Christians: modes and effects of Christianity among indigenous peoples* (Farnham)

VITEBSKY, P. (1993) *Dialogues with the Dead* (Cambridge)

VIVEIROS DE CASTRO, E. (1992) *From the Enemy's Point of View* (trans. C.V. Howard) (Chicago)

——(1996) 'Os pronomes cosmológicos e o perspectivismo Ameríndio', *MANA* 2.2: 115–44

——(1998) 'Cosmological Deixis and Amerindian Perspectivism', *Journal of the Royal Anthropological Institute* NS 4: 469–88

——(2004) 'Perspectival Anthropology and the Method of Controlled Equivocation', *Tipiti* 2.1: 3–22

——(2009) *Métaphysiques cannibales* (Paris)

VOLKOV, A. (1997) 'Zhao Youqin and his calculation of $\pi$', *Historia Mathematica* 24: 301–31

WAAL, F.B.M. de (1991) 'The chimpanzee's sense of social regularity and its relation to the human sense of justice', *American Behavioral Scientist* 34: 335–49

——(1996) *Good Natured* (Cambridge, Mass.)

WAGNER, R. (1975) *The Invention of Culture* (Englewood Cliffs, NJ)

——(1978) *Lethal Speech. Daribi Myth as Symbolic Obviation* (Ithaca, NY)

——(1991) 'The fractal person', in M. Godelier and M. Strathern (eds.) *Big Men and Great Men* (Cambridge), 159–73

WEISKRANTZ, L. (2009) *Blind Sight* (1st edn. 1986) (Oxford)

WILLERSLEV, R. (2007) *Soul Hunters* (Berkeley)

WILLIAMS, B.A.O. (2002) *Truth and truthfulness: an essay in genealogy* (Princeton)

ZIFF, P. (1960) *Semantic Analysis* (Ithaca, NY)

——(1972) *Understanding Understanding* (Ithaca, NY)

# INDEX